D0773190

The New Scientific Spirit

The New Scientific Spirit

Gaston Bachelard

Translated by Arthur Goldhammer

Foreword by Patrick A. Heelan

Beacon Press
Boston

108, boulevard Saint-Germain, 75006 France
Originally published as *Le nouvel esprit scientifique*

Beacon Press books are published under the auspices of the Unitarian Universalist Association of Congregations in North America, 25 Beacon Street, Boston, Massachusetts 02108
Published simultaneously in Canada by Fitzhenry and Whiteside Limited, Toronto.

Printed in the United States of America

(hardcover) 9 8 7 6 5 4 3 2 1

Library of Congress Cataloging in Publication Data

Bachelard, Gaston, 1884-1962.
The new scientific spirit.

Translation of: Le nouvel esprit scientifique.
Bibliography: p.
Includes index.
1. Science—Philosophy. 2. Physics—Philosophy.
I. Title.
Q175.B143 1985 501 84-14609
ISBN 0-8070-1500-8

Contents

Foreword

Patrick A. Heelan

Gaston Bachelard's writings[1] are comparable in spirit and scope to those of Charles Sanders Peirce, the American founder of pragmatism.[2] Each in his time generated a ma-

[1] Gaston Bachelard was born at Bar-sur-Aubes in 1884 and died in Paris in 1962. For a bibliography of his writings, see Jean Rummens, "*Gaston Bachelard: une bibliographie,*" *Revue Internationale de Philosophie,* vol. 17 (1963), 492-504; or the brief bibliography in Dominique Lecourt, *Marxism and Epistemology: Bachelard, Canguilhem, and Foucault,* trans. by Ben Brewster (London: New Left Press, 1975), 111-113.

[2] Cf. Charles Sanders Peirce, *The Collected Papers of Charles Sanders Peirce.* 8 vols., 1-6 ed. by Charles Hartshorne and Paul Weiss; 7-8 ed. by Arthur Burks (Cambridge, Mass.: Harvard University Press, 1931-1958).

jor rupture in philosophy of the sciences, Peirce at the turn of the century in Anglo-American philosophy, Bachelard in the period spanning the Second World War in French philosophy. Like Peirce, Bachelard was a practicing scientist — he was a physical chemist — and like him, it was the experience of science which drove his philosophical critique. This critique was directed in the first place against the writings of contemporaries, such as Emile Meyerson,[3] which, he said, imposed on science a kind of philosophical viewpoint which served the cultural interests of the past but which distorted actual science. This was Bachelard's critique of the cultural hegemony of the philosophical establishment since Descartes, the goal of which was to establish and vindicate itself as the sole legitimator and interpreter of the sciences.

To this critique, to which Richard Rorty in his *Philosophy and the Mirror of Nature* has returned,[4] Bachelard added a complex analysis of the epistemology of the sciences, arguing that the specialized sciences were sources of real epistemological novelty. Here, mathematical invention and instrumental artistry collaborated to make manifest — but only to skilled scientists — realities unknown to and unanticipated even in principle by traditional philosophy. It would be arrogant, he said, for philosophy to claim to legitimate such discoveries, for in his opinion it has no prior jurisdiction over the process nor over the physiognomy of the entities revealed. Philosophy's lesson is to learn from such

[3] Cf. Emile Meyerson, *Identity and Reality,* trans. by Kate Loewenberg (New York: Dover, 1962). Many of Meyerson's more important works have not yet been translated.

[4] Richard Rorty, *Philosophy and the Mirror of Nature* (Princeton: Princeton University Press, 1979).

discoveries that it itself is the source of unconscious bias ("epistemological obstacles") to new knowledge, that only the creative imagination can break through such obstacles. In making such "ruptures," the poet and the scientist — not the philosopher — are the expert practitioners.

Bachelard's naturalism — that we are not outside of the material world we know but a natural part of it — opposed the traditional Mind-Body dualism of French Cartesianism, as exemplified by the great Enlightenment thinkers, Denis Diderot (1714-1784); Etienne Bonnat, Abbé de Condillac (1715-1780); Marie Jean Caritat, Marquis de Condorcet (1743-1794); Pierre Simon, Marquis Laplace (1749-1827); Adrian Marie Legendre (1752-1833); Jacques d'Alembert (1717-1783); and others.[5] They took the Mind to be a realm apart from Nature, capable of receiving information from Nature and of forming a scientific image, guaranteed by the very structure of the Mind, capable of mirroring in itself the objective bodily source of its impressions, Nature. For them, Nature was a Book written in geometrical language to be deciphered in objective terms by the Mind whose own cultural or historical goals and interests were in no way to be reflected in the outcome.

Bachelard also opposed idealisms in the Kantian and Hegelian traditions, as exemplified, say, by Léon Brunschvicg (1869-1944) and Jean Hyppolite (1907-1968), which sought guarantees for scientific knowledge in the structure of the human spirit, or of Absolute Spirit, or in an alleged unity of scientific method — Claude Bernard's classic work would

[5] Cf., for example, Michael J. Morgan's *Molyneux's Question* (Cambridge: Cambridge University Press, 1977), which is a useful study of these classic authors.

have fallen under this critique[6] — rather than temporary rest in fallible and changing accommodations between the human organism and nature.[7] In this latter respect, he reflected the influence of Emile Boutroux (1845-1921).

For Bachelard, the conventionalism of Pierre Duhem (1861-1916) or Henri Poincaré (1854-1912) were likewise affected by the idealism of too literary a "reading" of science, as if the claims of science could be read simply from the literary and graphic media in which they are communicated to the "intellectual" public, apart from the experimental praxis or "phenomeno-technology" of the scientific community. For him, the norms for scientific communication should not be determined by the intellectual tradition of philosophy, but by the phenomeno-technology of science which has, he claims, the power to exhibit scientific phenomena through the use of appropriate instrumentation.

Bachelard's naturalism was in his own view "dialectical"; that is, it moved between mathematics and experiment; mathematics projecting a structure of entities to be looked for, experiment exhibiting these entities as found through phenomeno-technology. The dialectic then tunes the instrument to the mathematical theory and adjusts the mathematical theory to what can be exhibited successfully with the instrument.

This dialectical naturalism endeared Bachelard to one wing

[6] *An Introduction to the Study of Experimental Medicine*, trans. by H. C. Greene (New York: Dover, 1957), originally published in 1865. Cf. Paul O. Hirst, *Durkheim, Bernard, and Epistemology* (London: Routledge and Kegan Paul, 1975). Chapter 1 compares Bernard and Bachelard.

[7] For a comparison of Bachelard with French writers on the philosophy of science, cf. René Poirier, "*Autour de Bachelard épistemologique,*" in *Bachelard: Colloque de Cerisy* (Paris: Union Generale d'Editions, 1974).

of the French Marxists after the Second World War. Louis Althusser, for example, saw his views as supporting a genuinely "dialectical" and "materialist" account of scientific knowledge against those who would subordinate the knowledge claims of science and their interpretation to religious or other cultural ideologies. Bachelard, it must be remembered, attacked the role of ideologies — among which he counted traditional philosophy — as introducing "obstacles" into scientific research by presenting scientific theories to scientists and to the general public always in terms of some obsolete philosophy fashioned to favor the interests of — in this case — (European) intellectuals who belonged to the leisured class. In contrast, scientific inquiry, for Bachelard, itself "secretes" its own epistemology, which scientists need not and should not fear, and from which true philosophy can learn but only by attending to the praxis of science.

Bachelard's naturalism contained within it a polemic against positivism, of which the founder was Auguste Comte (1798-1857). In Bachelard's view, positivism — and empiricism, likewise — sought to impose yet another ideology on science, each thereby generating its own share of "epistemological obstacles" which the research scientist had to overcome in order to advance. In both cases, the ideology is that of simple natures, knowledge of which, like riches for the bourgeoisie, is acquired cumulatively. In contrast, for Bachelard, experience is not of simple natures, but of group theoretic relations, and these in turn are periodically transformed by new scientific praxes. Bachelard's opposition to empiricism and positivism was shared by French philosophers of science, such as Georges Canguilhem

(1904-   ), Jean Cavailles (1903-1944), and Jean Ladrière, who mostly were his students. Consequently, in the decades after World War II, with the exception of a few scholars, such as Paulette Fevrier and Jean Destouches (1894-1961), there was no counterpart in French philosophy of science to the logical empiricist movement in America. More recently, however, one finds French philosophers of science, often trained in two disciplines, such as Bernard d'Espagnat, René Thom, Roland Fraissé, O. Costa de Beuregard, and Francis Jacques, who are beginning to respond to Anglo-American work in the philosophy of science.

Anglo-American logical empiricism, however, has changed in many ways since Bachelard's death in the early sixties.[8] A number of Bachelardian themes — scientific observations are theory laden, ''revolutions'' occur in the history of science, science is not value neutral, science is a community endeavor and reflects community interests, among others — are all presently part of the confused picture of science in the new Anglo-American philosophy of science. Although Bachelard's writings considerably antedate these changes, there is little evidence that Bachelard's views were influential to any significant extent in bringing them about. His major works in the philosophy of science were not available in English and they were not written with attention to formal, logical, and technical matters of style in such a way as might win for them consideration by professional philosophers of science outside of France. The principal agents of change within Anglo-American philosophy of science were in fact the critical writings by, to mention just a few,

[8] Cf. Frederick Suppe, *The Structure of Scientific Theories* (Urbana, Illinois: University of Illinois Press, 1974).

Stephen Toulmin, N. Russell Hanson, Paul Feyerabend, Mary Hesse, Karl Popper, as well as those by the historians of science, influenced by Thomas S. Kuhn and by recent work in the sociology of science, particularly in Britain. These changes bring Bachelard's work into relevance for the mainstream of current philosophy of science in America and indicate the peculiar importance of some of his contributions — the role of imagination, epistemology as historical, ontology as the acceptance of a value, science as the effort to produce ''epistemological ruptures,'' the stress on instruments as constituting a ''phenomeno-technology.'' Bachelard's own idiom, once difficult to translate in a way that would make it accessible and intelligible to Anglo-American philosophers, is no longer so difficult, and Arthur Goldhammer has made an excellent translation, putting it into contemporary English.

State University of New York, Stony Brook
July 1984

Translator's Preface

Gaston Bachelard, born in 1884 in Bar-sur-Aube, France, the son and grandson of a cobbler, died in 1962 a member of the Institute and one of the most illustrious names in French philosophy. This book, first published in 1934 and currently in its fifteenth French edition, followed half a dozen earlier works by Bachelard in what for want of a better term I shall call the philosophy of science. Although Bachelard continued to write and publish in this area until the end of his life, the bulk of his writing after 1938, when the *Psychoanalysis of Fire* was first published, was concerned with matters of aesthetics and poetics, seemingly quite remote from theoretical physics and quantum chemistry. It is this latter part of Bachelard's work, much of which has already been translated into English, that is best known in this country.

In France, too, it was the poetics and "materialist psychoanalyses" that won Bachelard a wide audience, but he had previously attracted the attention of a more limited, but powerfully influential, group of academic readers with his work in what has come to be known as "historical epistemology." It was this interest that earned him, in 1940 at the age of 56, the invitation to come to Paris in order to take up the prestigious chair in the History and Philosophy of Science previously occupied by his teacher and patron, Abel Rey. Over the next twenty years Bachelard taught or otherwise influenced many of those now most prominent in French philosophy, most notably Georges Canguilhem, Jean Hyppolite, and Michel Foucault. But it was probably the influence of Louis Althusser that did most to revive the interest of the students of the 1960s in the early, epistemological, period of Bachelard's work. The reasons for this are beyond the scope of this brief preface.[1]

[1] See, among other works, Dominique Lecourt, *L'épistémologie historique de Gaston Bachelard* (Paris: Vrin, 1974). An English translation by Ben Brewster of this brief work, Lecourt's *mémoire de maîtrise*, is contained in Lecourt, *Marxism and Epistemology: Bachelard, Canguilhem, and Foucault* (London: New Left Books, 1975). It is hard to know what Bachelard would have made of Althusser's appropriation of his notion of *coupure épistémologique* to demonstrate a supposed discontinuity between the early, prescientific Marx and the later, scientific author of *Capital*. See Louis Althusser, *For Marx* (New York: Vintage, 1970). By all accounts Bachelard was a most genial man and would probably not have cried "foul," but he was also traditional enough to doubt whether even biology was a true science, and I dare say he would have been still more dubious of the claims

What kind of book is this, then, and why should anyone want to read it now, more than fifty years after its first publication?

The answer to the first question is not simple, because this is not a book written with a single purpose. It is, among other things, an impressive effort by a literate and learned man to appreciate and communicate what was novel about the "new physics" of his day, that is, relativity and quantum theory; it is also a sharp polemic against an influential tradition in philosophizing about science that Bachelard believed to be misguided and indeed pernicious; and it is an impassioned plea to free the teaching of science from fetters imposed by unreflective acceptance of that philosophical tradition.

Bachelard's appreciation of the new physics is not only warm but more importantly shrewd. To a degree rare among French philosophers of his day he understands the spirit of mathematics and not just the letter. He knows that the logic of discovery is not a matter of chaining together syllogisms of the "All men are mortal, Socrates is a man" variety. He sees that systems of axioms are formally prior but psychologically posterior to the "work" of the mathe-

---

of scientific socialism. For Bachelard as for the philosophical traditionalists he opposed, mathematics remained the "queen of sciences," and his "demarcation criterion" for a science, had he bothered himself about such a thing, would surely have been mathematization. But unlike Karl Popper, Bachelard seems not to have been concerned with legislating what ought and ought not to be considered science. Like Popper he had his doubts about Freudian psychoanalysis, but he was not a man to suppose that his criticisms could be given decisive force by "demonstrating" that psychoanalysis fails to satisfy some philosopher's definition of science.

matician and the theoretical physicist. He has grasped the fact that the "objects" that mathematicians and physicists investigate are not offered gratis by nature but have to be cajoled from it.

The only defect in his appreciation of science is that it is so allusive. Bachelard, in all of his writing, was able to make certain assumptions about who his audience was and what it knew. Here, for example, he assumes that his readers will be familiar with various topics of current research in physics and chemistry and with specific technical terminology. And there does indeed seem to have been a lively interest in current scientific research among French philosophers in the 1930s: A cursory survey of the *Revue de Métaphysique et de Morale,* where Bachelard published frequently and which was the only French journal I have been able to discover that reviewed this book when it first appeared, reveals articles by distinguished scientists and mathematicians such as Louis de Broglie and Jacques Hadamard. These were not much beyond the level of what one might find today in, say, *Scientific American,* but the spirit in which they are written is different. It is assumed without discussion that the results of science are of supreme interest to men of culture in general, and for the simple reason that it is better, ethically superior, to know the best that is known about the physical world before one speculates about anything else. The gap between physics and metaphysics is not assumed to be unbreachable. The philosophy of science, and even science itself, are not yet the private preserve of specialists. The general attitude is, in a word, neo-Kantian. Critique has not yet turned against science or the civilization that spawned it. Husserl's *Crisis* had not yet come to France.

Bachelard, as the reader will soon gather, dissents sharply in some respects from the neo-Kantian attitude, though most assuredly not from science or scientific civilization. For him the problem with all of the contending academic philosophies is that they are too prim, staid, fixed, and secondary. Philosophy takes a bookish knowledge of what science is, or, rather, has been, and turns it with Beethovenish insistence into a thunderous declaration: *Es muss sein.* René Poirier, in an interesting essay on Bachelard's predecessors in French philosophy of science, emphasizes the traits that they shared but that he did not: "their education, at the Ecole Normale, classical in nature, primarily based on Greek and Latin authors, and more or less Kantian as well . . . They had read the same traditional authors, most notably those of the seventeenth century; they thought in an orderly, methodical way, measuring their words. . . . All were born of bourgeois families and with Parisian souls."[2]

It is not a sin to be a mandarin, but the view from the heights may well strike someone more down-to-earth as needlessly remote. Bachelard, who before World War I had been employed as a part-time technician by the French Post and Telegraph Office, liked to get his hands dirty. The man who later wrote *The Psychoanalysis of Fire* and *Reveries of the Earth* had always been attracted to the blue flames and foul smells of the chemical laboratory. He delighted in setting up experiments for his pupils at the *collège* in Bar-sur-Aube.[3] It is no doubt an abusive extrapolation to suggest

[2] René Poirier, "Autour de Bachelard Epistémologue," in *Bachelard: Colloque de Cérisy* (Paris: Union Generale d'Editions, 1974), p. 11.

[3] "Table Ronde: Bachelard et l'enseignement," in *Bachelard: Colloque de Cérisy*, pp. 410–33.

that these biographical incidentals alone account for Bachelard's impatience with philosophical realism and phenomenology. Yet experience had taught him that our unaided powers of observation are small compared with the things we can see when we put our minds to work. Phenomenology is not enough, says Bachelard. We need a "phenomeno-technology," a way of producing experiences of the right kind. And whatever we build is informed by our prior view of how things are. All observations are, as we would now say, theory-laden.[4]

This was new at the time, and even newer was Bachelard's enthusiasm for the wondrous creative powers of the "new" mathematics. Jean Hyppolite has characterized Bachelard's philosophical work as "romanticism of the intellect," certainly an apt expression for his lyrical effusions about "non-Euclidean" mathematics. Prior to the present book Bachelard had published several works that bristle with the real stuff of mathematics, and it is rather odd that here he is at such pains to avoid symbols and equations altogether, even though their use would in many places have made his exposition more lucid. Bachelard's French can be translated, with difficulty, into English, but mathematics always translates badly into any language, and Bachelard's attempts to render equations into French are not, as it seems to me, always successful. I have added notes and parenthetical re-

[4] The term was introduced in English by N. R. Hanson, *Patterns of Discovery* (Cambridge: Harvard, 1959), but Bachelard much earlier raised doubts about the possibility of distinguishing sharply between theory and observation. For a recent critique of this idea from a writer who, like Bachelard, is sensitive to the complexities of experimental physics, see Ian Hacking, *Representing and Intervening* (Cambridge, England: Cambridge University Press, 1983), pp. 171ff.

marks where I thought these would be helpful but for the most have retained Bachelard's terminology, even where it now seems quaint (e.g., "microphysics"). To have done more would have been to write a new and different book from the one Bachelard wrote. Far too much more is known now, and any revision would suffer from a fatal loss of innocence. Anachronism is so alien to the Bachelardian spirit that even clarity must be allowed to suffer in order to avoid it.

The New Scientific Spirit is thus inevitably a historical document, and, I would argue, a historical document of a singularly important kind. The reason it should be read today — to answer the second question posed above — is that it provides an important clue to the origin and nature of a troubling divergence between what I shall call, with gross oversimplification, intellectual life here and in France. When we live the intellectual life, what are we about? Most of us, I think, would find this question very difficult to answer for ourselves but far easier to answer for our French *confrères*. The French "intellectual" (the word itself speaks volumes) is *carrying on a tradition*. This is so even for the thinker who styles himself a revolutionary, an iconoclast. Indeed, it goes a long way, I think, toward explaining why revolutionaries and iconoclasts are so common a feature of the French intellectual scene. The *ancien régime* has to survive in one way or another in order to be overthrown. Hence French thought, for all its bold and repeated affirmation of the dictum *Du passé faisons table rase* (Let's wipe the slate clean and start all over again), has for some time now been marked by a strong historicizing tendency. In the present work this takes the form of a two-pronged polemic, directed first against Descartes, at once the symbol and origin of the whole tradition, and second against the "establishment" in the philosophy of science that had developed since Claude Bernard.[5] Indeed, as Bachelard sees it philosophy itself is unserviceable, for in its traditional form it has always presumed to have legislative jurisdiction over science, and this is no longer tolerable. The productive forces of science have expanded to the point where this institutional-legal shackle on further development must be thrown off, as it were. Bachelard's critique of Descarte's famous meditation on the piece of wax is a *tour de force* whose brilliance will not, I hope, be lost on readers whose elementary educations did not necessarily include potted Cartesian wisdom. In any case, this book documents a key moment of change in the French philosophical tradition and can be read for the light is throws on the nature of and reasons for that change. This, in turn, acts as a *repoussoir* that helps to set our own position, and that of other European cultures, in proper perspective.

Bachelard's revolt against the philosophical establishment was of a relatively mild kind, however. One might compare it to a political revolution that briefly distracted attention from a slower but ultimately more devastating social revolution.

[5] Bernard is not an explicit target of Bachelard's polemic, but much of what he says can be construed as a criticism of Bernard's account of experimental medicine, which Bachelard would have regarded as too empirical a discipline to bear a full weight of a theory of knowledge. For a good comparison of Bachelard and Bernard (which, however, in my view understates the differences between them), see Paul Q. Hirst, *Durkheim, Bernard and Epistemology* (London: Routledge & Kegan Paul, 1975), chap. 1. See also Georges Canguilhem, "L'évolution du concept de méthode de Claude Bernard à Gaston Bachelard," *Etudes d'Histoire et de Philosophie des Sciences* (Paris: Vrin, 1975.)

It is interesting, in this connection, to recall the words of the young Raymond Aron in a review of, among other things, the book Bachelard published just after the present one, *The Dialectics of Duration: "Recherches Philosophiques,"* writes Aron, "has introduced many German doctrines to France, in French. We heartily approve this influence, which promises to have fruitful effects on a philosophy mired in its traditions and almost wholly absorbed by the theory of knowledge — epistemology — or literary discourse."[6] Clearly this infusion of German ideas had a far more profound effect on the future of French philosophy than Bachelard's domestic iconoclasm. Ironically, it was partly weariness of existentialism that turned French philosophers back to epistemology and hence to the work of Bachelard.

English-speaking readers will no doubt interpret Bachelard's place in the history of philosophy in their own way. There are surely anticipations in his work of many ideas that have gained prominence in recent Anglo-American debates on the philosophy of science.[7] But it is un-Bachelardian to look for precursors. A more salutary exercise, as he might say, would be to ask why our philosophizing didn't take

[6] Raymond Aron, book review, *Zeitschrift für Sozialforschung,* 1937, no. 6, pp. 417–20. It may be of some interest to point out that Aron, never one to pull his punches, found Bachelard's work in the book under review to be "brilliant and subtle yet [such as] to leave one with an unsettling impression. . . . The author, a specialist in the critique of the sciences, is probably uninterested in becoming familiar with philosophical method. Metaphysics provides him with an opportunity to embroider witty variations on his theme." This criticism, though that of a critic fundamentally out of sympathy with Bachelard's aims, is not inapt, and could equally well be applied to the present book.

[7] A good introduction is Frederick Suppe, ed., *The Structure of Scientific Theories,* 2d ed. (Urbana: University of Illinois Press, 1977).

a Bachelardian turn. What ''epistemological obstacles'' stood in the way (to borrow an expression from a later work with a title similar to that of the present book: *The Formation of the Scientific Spirit)*? But this is not the place to attempt an answer. It is far better to let Bachelard himself speak, to reveal what he saw clearly and what he did not. As a teacher he was beloved by his students, whose preconceptions he considered it his first duty to shake. It would be unfair to prepare you for his thrusts by giving away his strategy.

Thanks to Harry Marks of Harvard University for reading this translation and making helpful comments on the manuscript.

Arthur Goldhammer

Introduction

The Essential Complexity of the Philosophy of Science: An Outline

I

Since William James it has often been repeated that every cultivated man necessarily subscribes to some system of metaphysics. To my mind it is more accurate to say that every man who attempts to learn science makes use not of one but of two metaphysical systems. Both are natural and cogent, implicit rather than explicit, and tenacious in their persistence. And one contradicts the other. For convenience let us attach provisional names to the two fundamental philosophical attitudes that coexist so peacefully in the modern scientific mind: rationalism and realism, to use the classical terminology. Is proof required that such tranquil electicism does indeed exist? Consider, then, the following proposition: "Science is a product of the human mind, a product that conforms to both the laws of thought and the outside world. Hence it has two aspects, one subjective, the other objective; and both are equally necessary, for it is as impossible to alter the laws of the mind as it is to change the laws of the Universe."[1] This rather odd metaphysical assertion can be pursued in two possible directions: the first leading to a rationalism at one remove, according to which the laws of the universe would merely reflect the laws of the mind; the second leading to a universal realism, one of whose principles would be that the laws of the mind, being instances of universal laws, must be absolutely invariable.

The philosophy of science has done nothing to purify itself since Bouty enunciated the above proposition. It would not be difficult to show that, in forming scientific judgments, the most determined rationalist daily submits to the instruction of a reality whose ultimate structure eludes him, while the most uncompromising realist does not hesitate to make simplifying assumptions just as if he believed in the principles on which the rationalist position is based. One may as well admit that, as far as the philosophy of science is concerned, there is no such thing as absolute realism or absolute rationalism, and that judgments of scientific thought should not be couched in terms of general philosophical attitudes. Sooner or later scientific thought will become the central subject of philosophical controversy; science will show philosophers how to replace intuitive, immediate systems of met-

[1] Edmond Bouty, *La vérité scientifique* (1908), p. 7.

aphysics with systems whose principles are debatable and subject to experimental validation. What does it mean to say that science can "rectify" metaphysics? As an example of what I have in mind, consider how "realism" changes, losing its naive immediacy, in its encounter with scientific skepticism. Similarly, "rationalism" need not be a closed system; *a priori* assumptions are subject to change (witness the weakening of Euclid's postulates in non-Euclidean geometry, for example). It should therefore be of some interest to take a fresh approach to the philosophy of science, to examine the subject without preconceptions and free of the straitjacket imposed by the traditional vocabulary of philosophy. Science in effect creates philosophy. Philosophy must therefore modify its language if it is to reflect the subtlety and movement of contemporary thought. It must also respect the oddly ambiguous requirement that all scientific ideas be interpreted in both realistic and rationalistic terms. For that reason perhaps we ought to take as our first object of contemplation, our first fact needing explanation, the metaphysical confusion implicit in the double meaning of the phrase *scientific proof,* which can refer either to confirmation by experiment or to demonstration by logic, to palpable reality or to the mind that reasons.

It is fairly easy, moreover, to explain why any scientific philosophy must have such a dualistic base: The very fact that the philosophy of science is a philosophy that *applies* to another discipline means that it cannot preserve the unity and purity of speculative philosophy. Any work of science, no matter what its point of departure, cannot become fully convincing until it crosses the boundary between the theoretical and the experimental: *Experimentation must give way to argument, and argument must have recourse to experimentation.* Every application is a form of transcendence. I intend to show that this duality exists in even the simplest scientific investigations, that is, that the phenomenology of science divides, according to one set of epistemological polarities, into two realms, that of the picturesque and that of the comprehensible (which is just another way of saying that science may be viewed in either realistic or rationalistic terms). If we could somehow place ourselves at the frontiers of scientific knowledge and there observe the psychology of the scientific mind, we would find that it has been a concern of contemporary science to overcome the contradictions of metaphysics. Yet the orientation of the epistemological "vector" seems clear. It surely points from the rational to the real and not, as all philosophers from Aristotle to Bacon professed, from the real to the general. To put it another way, the application of scientific thought seems to me to tend essentially toward reality (*nous paraît essentiellement réalisante*). Accordingly, the purpose of this book will be to demonstrate what might be called the realization of the rational or, more generally, the realization of mathematics.

Furthermore, this need of application is felt just as strongly in pure mathematics, though there it is more hidden. It introduces an element of metaphysical duality into the mathematical sciences, which appear to be purely homogeneous, and thus offers a pretext for polemics between realists and nominalists. People seduced by the marvelous epistemological gain achieved by the introduction of formal axiomatic systems into mathematics (thereby enabling mathematical notions to function, as it were, in the void) may be too quick to condemn mathematical realism. Yet if one is careful not to abstract too hastily from the psychology of the mathematician, one quickly discovers that there is more to mathematics than formal structures, and that every pure idea is accompanied by an imagined application, an example that does duty for reality. In contemplating the work of the mathematician one also notices that such work always stems from an extension of knowledge derived from reality, and that within mathematics reality fulfills its true function: to provoke thought. In a reasonably clear-cut manner, mathematical realism (in its various functional roles) sooner or later operates to *give body* to pure thought (the French word is *corser,* which literally means to give body to wine by adding spirits — trans.); realism gives mathematical ideas psychological permanence; it parallels the spiritual activity, thereby disclosing (in mathematics as in other realms) the duality of subject and object.

Since the purpose of this book is chiefly to study the philosophy of the physical sciences, it is on the realization of the rational in physical experimentation that we must focus our attention. This realization, which has its counterpart in technological realism, is in my view one of the distinctive features of contemporary science, which in this respect differs markedly from the science of centuries past; in particular, it is quite remote from both the agnosticism of positivism and the tolerant attitudes of pragmatism and has nothing to do with traditional philosophical realism. It is rather a realism at one remove, conceived in reaction to the usual notion of reality, as a polemic against the immediate; it consists of realized reason, reason subject to experimentation. The "reality" to which this realism corresponds is not transferred into the realm of the unknowable thing-in-itself. It has a noumenal richness of quite another order. The *thing-in-itself* is a noumenon by exclusion of phenomena, whereas scientific reality, I would argue, consists in a noumenal context suitable for defining axes of experimentation. Scientific experiment is thus reason confirmed. This new philosophical view of science paves the way for the reintroduction of standards of experimental validity: Since the necessity of any given experiment is demonstrated in theory before being revealed by observation, the physicist's task is to purify phenomena sufficiently to recover the organic noumenon. Goblot has considered the use of proof by construction in mathematical argument; constructive arguments have lately made their appearance in mathematical and experimental physics. The whole doctrine of the "working hypothesis" seems to me destined to quick obsolescence. To the extent that hypotheses have been linked to experiment, they must be considered just as real as the experiments themselves. They are "realized." The time of the adaptable patchwork hypothesis is over, and so is the time of fixation on isolated experimental curiosities. Henceforth, hypothesis is synthesis.

If immediate reality has now become a mere pretext for scientific thought rather than an object of knowledge, it is time to move from descriptive *comment* to theoretical *commentary*. Prolix explication astonishes the philosopher, who would prefer to believe that explication is always limited to unfolding the complex, to demonstrating the simple within the composite. But true scientific thought is metaphysically inductive; as we shall see repeatedly, it reads the complex in the simple, states the law that covers the fact, the rule that applies to the example. We shall discover how modern scientific generalization, for all its vast scope, is the culmination of specialized knowledge. We shall uncover a kind of polemical generalization that shifts reason from the realm of the "why?" to the realm of the "why not?" We shall make room for paralogy alongside analogy and show how the ancient philosophy of the "as if" is superseded, in the philosophy of science, by the philosophy of the "why not?" As Nietzsche says, everything crucial comes into being only "in spite." This is as true of the sphere of thought as of the sphere of action. Every new truth comes into being in spite of the evidence; every new experience is acquired in spite of immediate experience.

Thus, apart from the gradual change in scientific thought brought about by the growth of knowledge, we will discover a virtually inexhaustible wellspring of novelty in the scientific spirit, novelty of an essential, metaphysical sort. Science is like a half-renovated city, wherein the new (the non-Euclidean, say) stands side by side with the old (the Euclidean). Anyone who thinks that such diametrically opposed idioms such as these are mere means of expression, more or less convenient systems of notation, attaches precious little importance to the proliferation of new scientific tongues. I, on the contrary, shall try to show that different means of expression may be more or less expressive, more or less suggestive, and therefore lead to more or less complete "realizations"; hence considerable importance must be attached to these expanded mathematical languages. Accordingly, I shall insist that such new doctrines as non-Euclidean geometries, non-Archimedean measures, non-Newtonian mechanics (associated with the name of Einstein), non-Maxwellian physics (associated with the name of Bohr), and noncommutative (or non-Pythagorean) arithmetic are valuable precisely because they confront us with suggestive dilemmas. Then, in the philosophical conclusion of the present work, I shall try to indicate the essential features of a non-Cartesian epistemology, to my mind capable of embracing all that is novel in contemporary scientific thought.

To avoid misunderstanding, it may be worth pointing out that there is nothing automatic about these negations, and one must not hope to find a simple formula for converting the new doctrines into terms comprehensible within the framework of the old. A genuine extension of knowledge has occurred. Non-Euclidean geometry was not invented in order to contradict Euclidean geometry. It is more in the nature of an adjunct, which makes possible an extension of the idea of geometry to its logical conclusion, subsuming Euclidean and non-Euclidean alike in an overarching "pan-geometry." First constructed in the margins of Euclidean geometry, non-Euclidean geometry sheds a revealing light on the limitations of its predecessor. The same may be said of all the new varieties of scientific thought, which have time and again pointed up gaps in earlier forms of knowledge. We shall discover that the new doctrines share many of the same characteristic features, the same methods of extension, inference, induction, generalization, complementarity, synthesis, and integration — all equivalents for the idea of novelty. And the novelty in question is profound: a novelty not of discovery but of method.

In the face of this ramification of epistemology, is there any justification for continuing to speak of a remote, opaque, monolithic, and irrational Reality? To do so is to overlook the fact that what science sees as real actually stands in a dialectical relationship with scientific reason. After centuries of dialogue between the World and the Spirit, mute experience is impossible. An experiment that purports to falsify the conclusions of a theory in a radical sense must justify its opposition to that theory. Physicists are not easily discouraged by negative experimental findings. Michelson, who believed in the existence of the ether, worked up to the moment of his death trying to refute the negative results of his own famous experiment.[2] Other physicists subtly reinterpreted Michelson's results by arguing that while negative in terms of Newton's system, they were positive in terms of Einstein's. Looked at another way, these physicists were merely applying the philosophy of "why not?" in the experimental sphere. A carefully done experiment is always positive, in other words. To say this is not to resurrect the absolute positivity of experimentation as such, for an experiment is not "careful" unless it is complete, that is, unless it is conducted according to a well-conceived plan based on a mature theory. Ultimately, experimental conditions are merely conditions of experimentation. What may seem a mere nuance of phraseology actually casts the philosophy of science in an entirely new light, by accentuating the technical difficulties involved in carrying through any preconceived theoretical project. The lessons of reality are valid only insofar as they suggest new ways to interpret theory.

[2] Albert A. Michelson (1852–1931), an American physicist who determined the speed of light to a high degree of accuracy. The "famous experiment" referred to in the text is usually known as the Michelson–Morley experiment, which failed to demonstrate any motion of the earth relative to the so-called ether, the material medium through which it was thought light propagates. — Trans.

Once one has meditated upon the nature of scientific action, it becomes clear that rationalism and realism are constantly exchanging counsels. Neither the one nor the other is in itself a sufficient philosophical basis of scientific proof. We cannot gain access to utlimate reality in the physical sciences through mere observation. Nor can any rational argument define the limits of experimental research once and for all. Hence there is a need for methodological innovation, a need that we shall have occasion to examine more closely later on; theory and experiment are so clearly related that no theoretical or experimental methodology is guaranteed to retain its validity indefinitely. To put the point even more strongly, even the most fecund of methods may eventually become sterile without the fertilizing stimulus of new problems to solve.

The epistemologist must therefore place himself at the crossroads between realism and rationalism. From this vantage he can grasp the new dynamism of those contradictory philosophies and study the dualistic process whereby science simplifies the real and complicates the rational. The gap between explicated reality and applied thought is reduced, thus circumscribing the area in which the philosopher must look to discover how standards of scientific proof are established (*la pédagogie de la preuve*). And such understanding is, as I shall point out in the final chapter of this book, the only possible psychology of the scientific mind.

More generally, it should be of some interest to examine the fundamental metaphysical question of the reality of the external world in terms of scientific practice. Why must philosophy always begin with the opposition between some vague Nature and some unformed Spirit, thus silently confounding the pedagogy of initiation with the psychology of culture? What audacity for the philosopher to assume that, abandoning his ego for a moment, he can recreate the entire World. And in any case what authorizes him to think that he can apprehend the simple, naked ego apart from its involvement in the acquisition of objective knowledge? In order to circumvent these basic questions, we shall look not only at the problems of science but also at the psychology of the scientific mind, regarding objectivity not as a primitive given but as something that is learned with great difficulty.

And there is more: It is in science, perhaps, that one sees most clearly the two meanings of the ideal of objectivity, the social as well as the concrete value of objectification. Science, as Lalande says, does not aim solely at "reconciling things with other things but even more at reconciling minds with other minds." Without the latter reconciliation, of course, there would be no problem. Faced with the most complex reality, we would, if left to our own devices, seek knowledge of a picturesque kind, calling upon our evocative powers: *The world would be our representation.* If, on the other hand, we were entirely given over to society, we would seek knowledge in the realm of the general, the useful, the conventional: *The world would be our convention.* In fact, however, scientific truth is a prediction or, better still, a predication. By announcing the scientific truth we call for a meeting of minds; together we convey both an idea and an experience, we link thought to experience in an act of verification: *The scientific world is therefore that which we verify.* Above the *subject* and beyond the *object,* modern science is based on the *project.* In scientific thought the subject's meditation upon the object always takes the form of a project.

Yet it would be a mistake to base one's argument on the rarity of true discovery in the history of man's Promethean efforts. For even in the humblest scientific thinking one finds evidence of the indispensable theoretical preparation for discovery. In an an earlier book [*La Valeur inductive de la relativité*] I did not hesitate to write that one does not point to (*montrer*) the real, one demonstrates it (*démontrer*). This is true particularly in cases involving an organic phenomenon of some kind (*organic* is here used in the broad sense of "having systematic structure" — trans.). When the object under study takes the form of a complex system of relations, then it can only be apprehended by adopting an appropriate variety of methods. Objectivity cannot be separated from the social aspects of proof. The only way to achieve objectivity is to set forth, in a discursive and detailed manner, a method of objectification.

It is of course a striking fact about science that it incorporates belief in demonstration as the basis of all objective knowledge. Observation is governed by a "code" of precautions that must be observed; observers are admonished to think before they look, to scrutinize carefully what they first see, and invariably to doubt the results of the initial observation. Scientific observation is always polemical; it either confirms or denies a prior thesis, a preexisting model, an observational protocol. It shows as it demonstrates; it establishes a hierarchy of appearances; it transcends the immediate; it reconstructs first its own models and then reality. And once the step is taken from observation to experimentation, the polemical character of knowledge stands out even more sharply. Now phenomena must be selected, filtered, purified, shaped by instruments; indeed, it may well be the instruments that produce the phenomenon in the first place. And instruments are nothing but theories materialized. The phenomena they produce bear the stamp of theory throughout.

The dialectical relationship between the scientific phenomenon and the scientific noumenon is not leisurely and remote but rapid and strict; after a few revisions, scientific projects always tend toward effective realization of the noumenon. A truly scientific phenomenology is therefore essentially a phenomeno-technology. Its purpose is to amplify what is revealed beyond appearance. It takes its instruction from construction. Wonderworking reason designs its own miracles. Science conjures up a world, by means not of magic immanent in reality but of rational impulse immanent in mind. The first achievement of the scientific spirit was to create reason in the image of the world; modern science has moved on to the project of constructing a world in the image of reason. Scientific work makes rational entities real, in the full sense of the word.

It is perhaps by examining the "technical activity" of thought that one can best gauge the essential philosophical dichotomy, which is summarized in Renouvier's "second metaphysical dilemma" under the head "dilemma of substance." The importance of this dilemma is crucial, for all the others follow from it. Renouvier puts it as follows: Either "substance is . . . a logical subject of undefinable qualities and relations'' or it ''is a being-in-itself and, as such, undefinable, unknowable.''[3] But it seems to me that experimental technology introduces a third term between the two horns of the dilemma, namely, the substantialized substantive. In general terms, the substantive, or logical subject, becomes substance once its qualities, regarded as a system, are unified in a role. We shall see later how scientific thought constructs entities whose unity derives from their function as key elements in a system. For example, a group of atoms in a synthesized organic substance can help us to understand the transition from ''logical'' chemistry to ''substantialist'' chemistry, that is, from the first horn of Renouvier's dilemma to the second. The dialectic of the physical sciences seems to me more instructive than the crude dialectic of traditional philosophy, because the poles between which it moves are less extreme, less heterogeneous. The study of scientific thought can actually afford the clearest insight into the psychological question of objectification.

II

To grasp the dialectic of contemporary scientific thought and show its essential novelty is the philosophical purpose of this little book. When one looks at science, what is immediately striking is that its oft-alleged unity has never been a stable condition, so that it is quite dangerous to assume a unitary epistemology. Not only does the history of science reveal a regular alternation between atomism and energetics, realism and positivism, continuity and discontinuity, ration-

[3] Charles Renouvier, *Les dilemmes de la Métaphysique pure* (Paris: F. Alcan, 1901), p. 248.

alism and empiricism; and not only is the psychology of the scientist engaged in active research dominated one day by the unity of scientific laws and the next by the diversity of things; but even more, science is divided, in actuality as well as in principle, in all of its aspects. Hence it has not been difficult to compile chapter after chapter illuminating this dichotomy. It would even be possible to reorganize the present work along somewhat different lines, so as to exhibit scientific reality as the intersection of two converging philosophical perspectives, the precision of theory always being subject to empirical correction. To determine whether or not a given chemical substance is pure, for example, one might specify its chemical function: The more clearly defined the function, the purer the substance.

Does the fact that the study of science forces such a dialectic upon us raise metaphysical difficulties for a philosophy that aims at synthesis? I have been unable to provide a clear answer to this question. In all controversial matters, I have tried to indicate the conditions that a synthetic view must satisfy, wherever reconciliation of an experimental or theoretical sort seemed possible. Yet in every such instance reconciliation seemed to me a compromise. A more important point, however, is the following: Reconciling contradictory points of view does not do away with the dualism that is inscribed in the history of science itself, in every conceivable approach to the teaching of science, and indeed in the very structure of thought. Apparent dualities in the immediate phenomenon might well be eradicable: They could be explained, for instance, as fleeting anomalies or momentary illusions and said to place the phenomenon in question in an entirely different category. But this way of proceeding proves unsatisfactory when the ambiguity is found to reside in the scientific phenomenon itself. What I am proposing, therefore, is a new way of looking at ambiguity, a view sufficiently supple to comprehend the recent teachings of science. The philosophy of science is, I believe, in need of genuinely new principles. One such principle is the idea that the characters of things may be essentially complementary, a sharp departure from the tacit (philosophical) belief that being always connotes unity. If being-in-itself is a principle that communicates itself to the mind (much as the concept of field defines the relation between a material particle and the space in which it moves), then being is nothing but a symbol of unity. What would be needed, then, would be an ontology of complementarity less sharply dialectical than the metaphysics of the contradictory.

III

Without pretending in any way to be laying the philosophical foundations of modern physics, I hope to suggest how common philosophical positions must be modified to accommodate reality as it is revealed in the scientific laboratory. Quite obviously the scientist can no longer be a realist or a rationalist in the manner of those philosophers who believed that it was perfectly possible to confront Being in either its outward prolixity or its inner unity. For the scientist Being can never be wholly comprehended by either experience or reason. It is therefore the task of epistemology to explain the ever-changing synthesis of reason and experience, even though achieving such a synthesis philosophically might appear to be a hopeless problem.

As for the plan of the remainder of the book, chapter 1 is devoted to a study of the origins of non-Euclidean geometry, focusing on the dialectical division of geometrical thinking and the subsequent synthesis. This chapter will be as brief as possible, since its only purpose is to show the dialectic of reason at work in its simplest and purest form.

In chapter 2 we continue the study of the dialectic by taking up the study of non-Newtonian mechanics.

In the following chapters we encounter questions that are less general in nature and at the same time more difficult to answer. We shall have occasion to consider the following series of dilemmas: matter and radiation, waves and particles, determinism and indeterminism.

We shall discover that the last of these pairs profoundly affects our concept of the real, which is made to seem strangely ambivalent. The question then arises whether Cartesian epistemology, based as it is entirely on appeals to simple ideas, is adequate to describe contemporary scientific thought. We shall find that the spirit of synthesis that animates modern science does not operate at the same depth as the Cartesian notion of *composition,* nor with the same degree of freedom. We shall then attempt to show that what is involved in the free and far-reaching synthesis of modern science is the same dialectic we discovered earlier in our study of non-Euclidean geometry. Hence this final chapter is entitled "Non-Cartesian Epistemology."

I shall take advantage of every opportunity to stress the innovative nature of the modern scientific spirit. Frequently innovation will be sufficiently apparent merely by comparing examples drawn from the physics of the eighteenth or nineteenth century with examples from the present century.

In this way we shall discover the incontestable novelty of contemporary physical science, both in detail and in the general structure of knowledge.

Chapter One

Dilemmas in the Philosophy of Geometry

In a brief and necessarily elementary chapter we cannot hope to follow the prodigious developments that have taken place in the philosophy of geometry over the course of the past century. The dialectical progress of thought is clearer in geometry than in any other science, however, and it is therefore important that we attempt to characterize the development of geometry in dialectical terms, that is, that we show first how the antithesis between Euclidean and non-Euclidean developed and second how these two "opposites" were ultimately combined in a new and higher synthesis. I am particularly concerned, moreover, to elucidate the psychological changes that each stage in this process required.

I shall therefore begin with a few remarks on the development of non-Euclidean geometry, in order to show how it "freed" rationalism by severing its psychological ties to a closed and immutable system of logic. I shall then look at the circumstances surrounding the synthesis of the Euclidean and the non-Euclidean. This will clarify the way in which different geometries are related to one another and lead to a discussion of a possibly unfamiliar mathematical object, the group.

Since groups are gradually finding application in mechanics and physics, our investigations will lead us to consider, from a highly abstract standpoint, the experimental and theoretical coherence of geometric thought. To me it seems that the epistemological issues raised by the use of non-Euclidean geometries in mathematical physics are quite different from the much older question of the relation between logic and the physical world. In this respect, Poincaré's "philosophical error" (which forms the subject of part three of this chapter) may be seen as providing a measure of the psychological reform that has been achieved by science in the past century.

I

Recall that prior to the last century's period of turmoil geometry enjoyed a long period of undisturbed unity. In the two thousand years after Euclid the quantity of geometrical knowledge undoubtedly increased, but the nature of geometrical thought remained fundamentally unchanged. Indeed, it might even be said that changeless geometry served as a paradigm for human reason itself. It was on the immutable architecture of geometry that Kant based his architectonics of reason. With the advent of non-Euclideanism, however, the only course open to Kantians was to argue that something in the nature of reason itself required the new geometry; in other words, the very principles of rationalism had to be extended and broadened. It is no doubt historically inaccurate to speak of a "Hegelian" philosophy of mathematics. Still, one cannot fail to be struck by the fact that dialectical tendencies appear at almost the same time in philosophy and in science. Human reason has a destiny of its own. In Halsted's words, "the discovery of non-Euclidean geometry around 1830 was inevitable." I therefore want to take a brief look at the mathematical work of the late eighteenth century that paved the way to that discovery. But I should first remind the reader that contemporaries were not aware at first of the nature of the epistemological problem involved.

D'Alembert in fact considered the parallel postulate, which Euclid *insisted* must be true, as a *theorem* in need of proof.[1] No one at this time doubted that such a theorem did indeed correspond to a truth, to a mathematical fact. To put it another way, all geometers of the late eighteenth century believed that parallels *exist.* The parallel postulate was justified by ordinary experience as well as by its indirect consequences. What was missing, and what struck mathematicians as scandalous, was that there was no known way

[1] Jean le Rond d'Alembert (1717–1783), French mathematician and philosopher.

of coordinating this simple theorem with the other known theorems of geometry. The existence of parallels was never questioned. Here again, premature realism betrays a profound misunderstanding of the true nature of the parallel problem.

This misunderstanding persisted even as the path to discovery was being opened up. Saccheri[2] and Lambert[3] in the eighteenth century and Taurinus[4] and de Tilly[5] in the nineteenth all regarded the parallel postulate as a theorem to prove, a truth to establish, a fact to justify. But with them an element of doubt makes its first appearance, albeit initially in the guise of method. For these mathematicians asked themselves what would happen if one were to drop or modify the notion of parallel. Their method went slightly beyond the familiar *reductio ad absurdum*. Lambert did more than just compile a list of bizarre theorems. For example, he noticed the effect of modifying the parallel postulate on the area of a triangle. Indeed, he was aware that a non-Euclidean argument could be pursued some distance without violating the requirements of logic, as his demonstration of the analogy between straight lines in the plane and great circles in the sphere shows. In both cases a series of theorems could be deduced one from the other. A logical "chain" could thus be pieced together, independent of the

[2] Gerolamo Saccheri (1667–1733), Jesuit priest, mathematician, and professor at the University of Pavia.

[3] Johann Heinrich Lambert (1728–1777), German physicist, mathematician, astronomer, and philosopher, author of *The Theory of Parallel Lines*, written in 1766 and published in 1786.

[4] Franz Adolf Taurinus (1794–1874), German mathematician and friend of Gauss.

[5] Joseph Marie de Tilly (1837-1906), Belgian mathematician.

actual nature of the links. In an even more precise manner, Taurinus noticed that "the great circles of the sphere have properties quite similar to those of straight lines in the plane, except for the property expressed in Euclid's sixth postulate, namely, that two straight lines can never enclose a space (i.e., a region of positive area)."[6] Taurinus's formulation in terms of "space" was frequently taken to be equivalent to the classical parallel postulate.

These simple remarks — very early manifestations of the non-Euclidean spirit in geometry — are enough to indicate the general nature of the new freedom of mathematics. Already it is apparent that the *role* played by an object is more important than its *nature,* and that essence coincides with relation. In other words, mathematicians had begun to take the attitude that one cannot understand the issues raised by Euclid's postulate until one has truly understood, not the "absolute" or "essential" nature of lines in the plane, but the role they play in making logical arguments; and that one will understand this role only when one has learned to generalize the function of lines in other applications, that is, how to use the idea of a line outside the Euclidean domain in which it originated. Accordingly, simplicity is no longer, as in Cartesian epistemology, the intrinsic quality of an idea but merely an extrinsic and relative property, dependent upon the idea's being perceived in a particular relation deriving from a particular application. To put it in somewhat paradoxical terms, non-Euclidean geometry began with the purification of a pure idea, with the simplification of a simple notion. For Taurinus's observation leads

[6] Barbarin, *La géométrie non-euclidienne,* 3d ed., p. 8.

rapidly to the question whether straight lines that *do* satisfy the parallel postulate are not unduly restricted special cases, or, in other words, composite notions, since from the functional standpoint the great circles (the spherical analogues of the lines in the plane) do not satisfy the postulate. This is precisely what Barbarin has in mind when he points out that as early as 1826 Taurinus had offered his opinion that "if Euclid's fifth postulate is not true, the reason is probably that there are curved surfaces on which certain curved lines have properties analogous to those of straight lines in the plane, except for the property stated in the fifth postulate, a bold guess that would be proved right forty years later when Beltrami discovered the pseudosphere."[7] Later, when mathematicians finally came to regard straight lines as the geodesics of the plane, they were in fact harking back to Taurinus's guiding insights, which were to emphasize "extension" over "intension" and to think of mathematical objects only in terms of rigorously defined functional roles.

One must be careful not to assume too hastily, as a realist would, that it is now simply the surface rather than the line that is regarded as real, and that it is merely the fact of belonging to a surface that gives reality to the line. The question of realism in mathematics is more obscure than this simple view of the matter would suggest and cannot be approached in such a direct, straightforward, and concrete manner. It would be more nearly correct to say that the greater the variety of surfaces to which a line belongs, the more real it is; or, better still, that the measure of realism in mathematics is based on extension rather than intension.

[7] Ibid., p. 7.

In general, material reality may be defined in terms of anything that remains invariant over a sufficiently broad range of applications. The same is true of mathematical reality. Let me emphasize one point: The "reality" of a mathematical idea is a function of its extension and not of its intension — the geodesic is "more real" than the straight line. Mathematical thought comes into its own with the appearance of such ideas as transformation, correspondence, and varied application. It is precisely in a dialectical process that generality attains its maximum, moreover, as the most alien forms are unified by transformation. It is through such a process that mind may measure its grasp of mathematical reality. Let us now proceed, therefore, to examine what was decisive about the non-Euclidean revolution.

The dialectic implicit in the constructions of Lobachevski[8] and Bolyai[9] is more readily apparent than in Lambert's work, for the chain of theorems that could be derived from non-Euclidean assumptions about parallels had continued to grow and gradually freed itself from the need to rely for progress on analogies with the Euclidean case. It would not be inaccurate to say that Lobachevski worked for twenty-five years not to discover his geometry but rather to generalize it. Indeed, it could not have been discovered in any other way. It is as if Lobachevski tried to prove that motion exists by getting up and going somewhere. Is it possible that he believed he would ultimately encounter a contradiction simply by pursuing long enough the process of drawing all possible deductions from an assumption that might be charac-

[8] Nikolai Ivanovich Lobachevski (1793–1856), Russian mathematician.
[9] Farkas Bolyai (1775–1856), Hungarian mathematician.

terized, at first glance, as absurd? This question raises innumerable further questions about the way in which psychology impinges on epistemology.

It is customary to look at the problem from a purely epistemological standpoint and to describe the origins of non-Euclidean geometry in roughly the following terms. Since it had turned out to be difficult to give a direct proof of Euclid's postulate, it made sense to attempt a proof by contradiction: to assume, in other words, that the parallel postulate is false and see what follows. Inevitably, it was thought, a contradiction would turn up. Then, since the logic was correct, one of the hypotheses had to be false. On this view of the matter, the purpose of the exercise was to reestablish the validity of the parallel postulate.

This epistemological account begins to look inadequate, however, as soon as one glances at Lobachevski's work. Not only is no contradiction revealed, but the reader quickly gains the impression that the whole question remained *open* in Lobachevski's mind. Unlike a proof by contradiction, which normally leads rapidly to an absurdity, the deductions turned up by the Lobachevskian dialectic estabish themselves ever more firmly in the reader's mind. Psychologically speaking, there is no more reason to expect a contradiction in Lobachevski's work than in Euclid's. Such an equivalence would in fact be established later, in a technical sense, by the work of Klein[10] and Poincaré;[11] but its psychological influence is perceptible at a much earlier date. The difference here is a subtle one, invariably neglected by philosophers who base their judgments of what happened on

[10] Felix Klein (1849–1925), German mathematician.
[11] Jules Henri Poincaré (1854–1912), French mathematician.

the achievement of definitive results. Yet anyone who hopes to penetrate the new dialectic of science must develop psychological empathy with the participants, and the best way to do this is to study the whole spectrum of complementary ideas surrounding any discovery, in the moment of their first formation.

In short, any psychologist of the scientific spirit must actually experience for himself the peculiar *splitting of the geometric personality* that occurred in nineteenth-century mathematics. This is the only way to appreciate why the rather skeptical account given by ''conventionalist'' philosophers offers at best a pale version of the actual history of geometry, animated as it was in the past century by the stormy dialectic of Euclidean and non-Euclidean.[12]

Once one has appreciated the nature of this dialectic, it is only natural that one should see important general concepts of mathematics in a new light. In a letter to de Tilly written in 1870, Houel[13] characterized the greater generality of the new approach to geometry by an ingenious analytical comparison:

> The Euclideans were convinced that we were denying their geometry when in fact we were only generalizing it, since Lobachevski and Euclid can get on quite well together. Generalized geometry . . . is a method analogous to that employed by an analyst

[12] Conventionalism, the philosophical doctrine that all principles, axioms, postulates, and so on, are mere conventions chosen to suit our convenience, was fashionable in France around the turn of the century; Poincaré was one of its principal exponents. — Trans.

[13] Guillaume-Jules Houel (1823–1886), French mathematician.

> who, having found the general integral of the differential equation associated with some problem, will proceed to analyze that integral before selecting the constant of integration necessary to give the solution to the problem in question, though of course without any intention of denying that the arbitrary constant must ultimately be assigned some definite value. As for all the backward Euclideans who are still seeking to demonstrate the Postulatum, perhaps the best characterization I can give of them is to say that they resemble someone who tries to determine the constant of integration from the differential equation itself.[14]

This excellent comparison gives an idea of the synthetic role played by a set of axioms: A differential equation is obtained by eliminating arbitrary constants; its general integral subsumes all possible solutions of the equation; similarly, pangeometry eliminates arbitrary assumptions or, rather, neutralizes their effects by explicitly stating what is being assumed. The axioms are a distillation of complementary possibilities. Euclidean geometry then becomes one member of a set of possibilities, a special case.

The existence of many geometries helps to make each particular geometry less concrete (*déconcretiser*). Realism is a property not of any one geometry but of the whole set. Now that we have seen how the dialectic produced from the one Euclidean geometry a plethora of new geometries, we must go on to investigate how these were linked together in a coherent new synthesis.

[14] See the *Bulletin des sciences mathématiques*, February 1926, p. 53.

II

To understand the new mathematical realism we must understand the post-Euclidean synthesis, and we cannot do this by studying Euclidean geometry alone. Rather, we must try to discover what different geometries have in common, how one geometry corresponds to another. Mathematics takes on reality by establishing correspondences. Mathematical objects are known through their transformations. To the mathematical object we may say, Tell me how you are transformed and I will tell you what you are. Indeed, the way in which mathematicians prove that "different" geometries are actually equivalent is to show that they "correspond" to the same algebraic objects ("invariant forms").[15] Once a correspondence of this kind had been established, the fear of eventually finding contradictions in exotic geometries was dispelled: It was no more to be feared that a contradiction would be found in Lobachevski's geometry than in Euclid's, since any geometric contradiction would be reflected in the algebraic form and hence in all the geometries in correspondence. The cornerstone of the whole edifice, then, is the algebraic form. Put simply, algebra contains all relations and nothing but relations. The "equivalence" of different geometries is defined in terms of relations, and it is as relations that geometries have reality, not by reference to any object, experience, or intuition. Next, therefore, I want to examine, first, how the basic notions of geometry (such as

[15] A *form* is a rational, integral, homogeneous function of a set of variables; an *invariant form* is a form whose value remains essentially unaltered when its component variables are subjected to transformation by the elements of some group. Klein (see n. 10) uses invariant forms extensively in his work on geometry. — Trans.

point and line) were stripped of their vivid, concrete qualities and, second, how that vanished concreteness was shifted to the relations that are in fact the defining elements of the new geometries.

Consider, to begin with, Juvet's penetrating remarks on axiomatic systems.[16] Juvet points out that physics starts with notions quite remote from immediate experience; he further shows how theorists, far from enriching these basic notions with intuitive insights, have gradually purged and simplified them still further. Physical theories achieve maturity by reducing the intension of their constituent notions to the set of attributes visible in the extension of those notions. "By progressively paring away superfluous attributes it has been possible to avoid the kinds of antinomies that inevitably arise when the intension of a notion is too broad." In geometry this paring process was carried so far that it was once proposed that all empirical referents be proscribed from the subject. Juvet recalls the starting point of Hilbert's axioms: "Let there be three categories of objects, and call members of the first category A, B, C . . . members of the second category a, b, c . . . and members of the third category α, β, γ . . . It turns out later that the capital letters stand for the points, the small letters for the lines, and the Greek letters for the planes of elementary geometry."[17] Thus every possible precaution was taken to ensure that the objects of the theory should be understood, so to speak, "from above" rather than "from below" as when their origins lay in the world of substance. To put it still another way, the properties of

[16] Gustave Juvet, *La structure des nouvelles théories physiques* (Paris: F. Alcan, 1933), p. 157.
[17] Ibid., p. 158.

the objects in Hilbert's system are purely relational and in no way substantial.

But if the roots of the relations among the objects do not lie in the objects themselves, if the objects acquire their properties only *after* relations are imposed upon them, then we must be very careful about asking where these relations come from. Here again contingency reigns supreme, since the constitutive relations of a geometry must be absolutely independent of one another, and it must be possible to replace any axiom by its negative. We cannot arbitrarily declare any particular relation to be "real" once we rule out the possibility of appealing to some "substantive reality" as our reason for preferring such and such an axiom to its contrary. But if we have a coherent set of relations, things become less arbitrary; we can try, for instance, to add further relations to that set in order to make it "complete." In this way we work toward a "synthesis." And once we have found a complete set of relations, the geometry acquires a new coherence (*totalité*); it assumes an objective and not merely a notional unity. This, then, is the point at which the mathematical reality emerges. This reality does not coincide with the "primitive objects" of the theory, nor does it coincide with the relations among the objects taken one by one. Epistemologically, the crucial step in constructing a theoretical counterpart to reality (*la fonction epistémologique essentielle à toute réalisation*) comes when a new relation has to be added to an already large set of existing relations.

In what, precisely, does belief in reality consist? What is the idea of reality? What is the primordial metaphysical function of the real? My answer is this: The belief in reality is essentially the conviction that an entity transcends immediate sense data; or, to put the same point more plainly, it is the conviction that what is real but hidden has more content than what is given and obvious. Of course it is in the realm of mathematics that this "realizing" function operates most delicately. It is here that realization is most difficult to discern yet most instructive to apprehend. Let us therefore take Hilbert's nominalism as our starting point. Let us accept, provisionally, the requirements of absolute formalism. Let us erase from our memories all the beautiful objects of geometry, all those lovely forms, and henceforth think only of letters, not things! Then submit ourselves to absolute conventionalism: All the clear relations of geometry are now mere syllables that associate with one another in some mysterious way! What we have, then, is all mathematics, summarized, symbolized, and purified! This, however, is where the creative, poetic, "realizing" labor of the mathematician comes in: Suddenly, in a revealing turn, the associated syllables form a word, a real word, that both speaks to Reason and evokes some thing in Reality. This sudden acquisition of semantic value is an all or nothing phenomenon. The word takes its meaning from the completed sentence, not from the etymological root. It is when a notion is exhibited as a totality that it plays the role of something real. After reading a few pages of Peano's work, Poincaré complained that he didn't understand Peanian. This was because he was reading it literally, in terms of isolated conventions, like a dictionary, without any real desire to put it to use. For as soon as one *applies* Peano's dicta, it becomes clear that they mirror thought, that they engage the mind and subject it to rules; and yet it is not at all clear where they acquire this psychological force, for the dialectic of form and substance proceeds at a much deeper level of thought than is generally believed. Yet it is certain that this psychological force exists. The poetic transcendence of Peanian would no doubt be difficult to explore if we had no common exposure to mathematical thinking. Juvet rightly points out that "when we lay down a set of axioms, we try to appear as though we are not making use of what the science whose foundations we purport to establish has already learned, but in fact it is only for things that are known that we try to establish an axiomatic basis."[18] Still, it is true that we find evidence of a characteristic doubleness in the new mathematics. Axiomatization now *accompanies* scientific development. The accompaniment may have been written after the melody, but the modern mathematician plays with two hands. And what he plays is utterly new; it requires diverse levels of consciousness, and an affected, yet active, unconscious. It is oversimplifying matters to repeat the commonplace that the mathematician knows not whereof he speaks. In fact, he pretends not to know. He *represses* his intuition. He *sublimates* his experience. Euclideanism remains the naive foundation for all generalization. As Buhl observes, "it is quite remarkable that merely by slightly extending certain aspects of Euclidean geometry one can give rise to a far more general geometry and even geometries."[19] In one sense, mathematics *is* generalization: It is an aspiration to completeness. What is complete is coherent and mature: Completeness is a sign that the process of objectification has reached its term.

[18] Ibid., p. 162.
[19] Adolphe Buhl, "Notes sur la géométrie non-euclidienne," in Paul Barbarin, *La géométrie non-euclidienne* (Paris: Gauthier-Villars, 1928), p. 116.

If a set of axioms is like the blueprint of a geometry, that blueprint is itself the image of something more profound, something closer to the source of mathematical inspiration: the group.[20] Every geometry (and more generally, I suspect, every mathematical organization of experience) has its own characteristic group of transformations. Again we see that mathematical existence is governed by rules of transformation. In the case of Euclidean geometry, this characteristic group is particularly clear and simple — so clear and simple, in fact, that its theoretical and empirical importance is not immediately apparent. I am speaking, of course, of the group of displacements.[21] For the congruence of two figures is defined in terms of displacements, and congruence is clearly the basis of metric geometry: Two figures are said to be congruent when by displacement one can be superim-

[20] Bachelard uses the term *group* in a somewhat archaic sense (see Felix Klein, *The Icosahedron*, trans. G. G. Morrice [New York: Dover, 1956], p. 5), according to which a group is a set G together with an operation * defined on G such that if a, b are elements of G. then a*b is also an element of G. We say, then, that G is *closed* under the operation *. A stronger definition of group is now customary, and all the groups that Bachelard considers are in fact groups in this stronger sense (he had not yet freed himself from the concrete origins of the group idea to the extent that he believed). The stronger definition requires that G have an identity element e such that $e*a = a*e = a$ for all a in G, and, further, that the operation * be associative, that is, that $(a*b)*c = a*(b*c)$ for all a, b, and c in G, and invertible, that is, for every a in G, there is an element b such that $ab = e$ (the identity element). Notice that a*b need not equal b*a; a group for which $a*b = b*a$ for all a, b in G is termed *Abelian*. The theory of groups has been considerably developed since Bachelard wrote this book. See I. N. Herstein, *Topics in Algebra* (New York: Blaisdell, 1964). — Trans.

[21] A displacement in the plane is either a translation (i.e., a linear movement in a specific direction), a rigid rotation of the plane around a fixed point, or a combination of the two. The set of such operations forms a group, where if a and b are displacements, a*b (see n. 20) is taken to be the composite displacement. — Trans.

posed precisely upon the other. Two displacements can obviously be combined to yield a third displacement, which we call the "product" of the first two. Hence any (finite) series of displacements can also be combined to yield a single displacement. This is the basic property that makes the set of displacements into a group.[22]

Is this an empirical or a theoretical fact? The striking thing is that such a question can even be asked, that the group turns out to be so central to the dialectic of theory and practice. It is for this reason that the group, or at any rate the idea of composing operations that is fundamental to its definition, has become the common basis of theoretical and experimental physics. By incorporating the group as one of its fundamental ideas, mathematical physics has demonstrated the primacy of theory. The point is best understood by thinking about the earliest form of mathematical physics, Euclidean geometry. Juvet puts it quite well: "Experience shows . . . that displacements do not alter figures; but the axiomatic theory *proves* this fundamental proposition."[23] Proof takes precedence over observation.

Until a group has been associated with a given set of axioms, one cannot be sure that that set of axioms is truly *complete*. To quote Juvet again, "if a group is represented by a geometry, the axioms of that geometry are noncontradictory, insofar as one is willing to accept the theorems of analysis. Furthermore, the axioms of a geometry are not complete unless that geometry is the exact representation

[22] Here, Bachelard is clearly using the weaker definition of *group* (n. 20); but the displacements are a group in the stronger, more modern sense as well, as the reader may verify. — Trans.
[23] Juvet, *La structure*, p. 161.

of a group; until one has found the group that is the rational basis of the geometry, the latter is incomplete and possibly contradictory."[24] In other words, the group furnishes proof that the mathematical object in question is "closed" (or self-contained, in a precise mathematical sense[25] — trans.). The discovery of a group brings to a close the era of more or less independent, more or less coherent conventions.

Meyerson has brilliantly argued that certain principles of permanence underlie all physical phenomena.[26] I agree but would add that the value of those principles is not realistic but theoretical: Myerson's "permanences" are actually the invariants of certain groups of transformations. In any case, it is in the determination of invariants that the mathematization of the real finds its true justification; the mathematics yields a set of "organic permanences" (*permanences organiques*). Once again I cite Juvet: "In the torrent of phenomena, amid the constant movement of reality, the physicist discerns things that endure; to describe them, his mind constructs geometries, kinematics, and mechanical models, whose axiomatization has as its purpose to make clear . . . what for want of a better term I shall call the useful intension of the various concepts that experience and observation have suggested to construct. If the system of axioms so constructed is the representation of a group whose invariants can be interpreted as corresponding to those enduring objects that experience has discovered in reality, then the physical theory is free from contradiction and an image of reality." Juvet also draws a parallel between work on

[24] Ibid., p. 169.
[25] See n. 20 above.
[26] Emile Meyerson (1859–1933), French philosopher.

groups and Curie's work on symmetries. He concludes that what we have here is both a method and an explanation.

III

It will by now be apparent that the abstract geometries established by different sets of axioms and their corresponding groups yield different kinds of mathematical physics, and that we must look to the groups in order to discover the precise relations between one physics and another. In particular, the supremacy of Euclidean geometry is no more assured than that of its characteristic group. And in fact this group turns out to be a relatively impoverished one. It has therefore relinquished its prestigious position to more richly structured groups, which lend themselves better to the job of providing a detailed theoretical account of sophisticated experimental results. This is why Poincaré's opinion that Euclidean geometry has in its favor the virtue of supreme convenience is no longer held seriously by anyone. Poincaré's view strikes me as more than a little inaccurate, and meditation upon its inadequacy offers more than one reason for prudence in forecasting the fate of human reason.[27] One reason for dwelling on this error is that putting it right is a good way to revise the values on which we base our theoretical judgments; at the same time it affords a good illustration of the importance of abstraction in contemporary physics. I shall therefore recall Poincaré's argument very briefly and then try to say just what is new — in one respect, at any rate — about the new epistemology.

[27] Cf. Emile Meyerson, *Le cheminement de la pensée* (Paris: F. Alcan 1931), vol. 1, p. 69.

After Poincaré had shown that various geometries are logically equivalent, he asserted that Euclid's would always remain the most convenient, and that if this geometry should ever come into conflict with physical experience, people would always prefer modifying the physical theory to changing the elementary geometry. Gauss, for example, had proposed an astronomical experiment to verify a theorem of non-Euclidean geometry:[28] He asked whether a triangle with stars for vertices, hence of enormous surface area, would exhibit the decrease in area predicted by Lobachevski's geometry. Poincaré did not agree that the results of such an experiment would be crucial. If the experiment succeeded, he maintained, some physicists would immediately conclude that the light rays had somehow been perturbed so that they no longer propagated along straight lines. In one way or another, Euclidean geometry would be saved.

In a subsequent chapter on non-Cartesian epistemology, I shall argue that what we have here is a case of trying to explain away an anomaly by introducing a perturbation, the reason for which is assumed *a priori* to be clear. In broad terms, the strategy is to argue that what is intellectually perspicuous is fixed for all time. The underlying assumption is that the clearest thoughts are always the first to emerge, and that they must therefore fix the frame of reference for all future research. If this epistemological hypothesis were correct, what would the methods of physical science be like? Experimental design would be based on Euclidean assumptions. Physical objects would be assumed to be solids and other possibilities ignored (e.g., the Lorentz transforma-

[28] Karl Friedrich Gauss (1777–1855), one of the greatest mathematicians of all time.

tions). The scientist's work would not be truly rational but rather based on acquired *habits* of "rationality." In other words, thought would be governed by a vast Euclidean infrastructure shaped by everyday experience of natural and manufactured solids. Any anomalies in the experimental results would then be explained by assuming that this *unconsciously Euclidean* view of things is correct. Gonseth puts it quite well: "Errors and corrections are determined with the — generally unconscious — intention of minimizing any discrepancy between the measured results and Euclidean geometry."[29]

But is this Euclidean structure, once thought characteristic of human intelligence itself, really definitive? It is now possible to answer this question in the negative, since contemporary physics is in fact organized around non-Euclidean models. In order for the bonds of Euclideanism finally to be broken, physicists had to confront a new set of problems with total independence of mind, after liberating themselves from Euclidean presuppositions by engaging in what one might call a kind of psychoanalysis. The new problems came from the field of microphysics. We shall see presently that microphysical epistemology is not "thing-oriented" (*chosiste*). Here I shall say only that in microphysics the basic object is not a solid. Indeed, it is no longer possible to regard the electrically charged particles of which all matter is composed as true solids. Even this negative assertion must not be interpreted in realistic terms, for otherwise it is no more valid than the realist atomism it is intended to refute. The modern physicist can provide a very deep proof of the con-

[29] Ferdinard Gonseth, *Les fondements des mathématiques* (Paris: A. Blanchard, 1926), p. 101.

tention that elementary particles are not solids, and the nature of this proof is quite characteristic of the new physics: Elementary particles are not solids because motion distorts their shape. The way to look at the question is in terms of a *mathematical transformation,* the so-called Lorentz transformation, which is incompatible with the displacement group associated with Euclidean geometry.[30] No doubt those still convinced that the Euclidean outlook is correct will try to interpret the physics of charged particles in terms of the traditional geometry, by imagining some special kind of contraction. But to do so is to engage in a pointless and indeed a dangerous exercise, because it is impossible to imagine in any clear way how a solid object can contract. It is better to turn the argument from clarity on its head and to judge things from outside, as it were, accepting the mathematical imperatives implicit in the fundamental group. Microphysicists no longer work with the common-sense notion of a nondeformable solid. Instead, they try to conceptualize the behavior of particles directly in terms of the Lorentz transformations. For them the Euclidean interpretation of phenomena is merely a simplified, not a simple, image of reality. It is easy for the microphysicist to see what is left out or neglected in this simplified picture. Psychologically, the modern physicist is aware that the rational habits acquired from immediate knowledge and practical activity are crippling impediments of mind that must be overcome in order to regain the unfettered movement of discovery.

Even if we continue to grant some weight to the argument of convenience, it must be said that when it comes to inter-

[30] Named after Hendrik A. Lorentz (1853–1928), Dutch physicist.

preting experimental results of the sort common in microphysics, Riemannian geometry is frequently the most convenient, economical, and clear mathematical tool.[31] But the debate should really be judged in abstract terms. What is involved is not a question of two languages or two images, much less two spatial realities; the issue, rather, is which of two levels of abstraction, two different systems of rationality, two methods of research is preferable. From now on, the group is the guiding principle of all theory. Experimental results can also be coordinated with groups.[32] This gives an idea of the power of mathematics to create reality (*la valeur réalisante de l'idée mathématique*). The old dialectic of Euclidean and non-Euclidean thus has repercussions at the deepest levels of experimental physics. The whole problem of scientific knowledge of the real turns on the initial choice of mathematics. When one has fully comprehended what for example Gonseth's work shows, namely, that experimentation is always dependent on some prior intellectual construct, then it is obvious why one should look to the abstract for proof of the coherence of the concrete. Empirical possibilities are in one-to-one correspondence with sets of axioms.

Here I shall end the present chapter. By reexperiencing the birth of non-Euclidean geometry, which marks the first appearance of alternative sets of axioms in the history of mathematics, we have acquired the necessary background

[31] Named for Georg Friedrich Bernhard Riemann (1826–1866), German mathematician.

[32] Presumably this refers to the possibility of using the Lorentz transformations to coordinate observations involving more than one frame of reference; the statement is rather ambiguous. — Trans.

in physico-mathematical culture to understand the material to be presented in what follows.

Chapter Two

Non-Newtonian Mechanics

I

Some years ago I wrote a book [*La valeur inductive de la relativité,* 1929] devoted to explaining in what sense the theory of relativity was truly and essentially new. In that book I laid stress especially on the inductive value of the new mathematics, showing in particular how the tensor calculus played the role of a veritable new organon. In the present chapter I have deliberately chosen to avoid mathematical details and will limit myself to a general comparison of the Newtonian with the Einsteinian scientific spirit.

Astronomy has been totally recast since Einstein. Relativistic astronomy is in no sense a child of Newtonian astronomy. Newton's system was complete, finished. By making minor adjustments to the law of gravitational attraction and perfecting the theory of perturbations, physicists were able to account in several ways for the precession of Mercury's perihelion and other anomalies. In this sense there was no need for a wholesale revision of the theory of gravitation simply to account for the observed data. What is more, the Newtonian world was a spacious and lustrous abode. Newtonian physics was from the first a marvelously clear example of a closed system of thought. The only way out was by force.

It is a mistake, in my view, to regard Newtonian physics as a first approximation to Einsteinian physics (simply because it gives roughly the same results for speeds much smaller than the speed of light — trans.), because the subtleties of relativity have nothing to do with any sophisticated application of Newtonian principles. It is therefore incorrect to say that the Newtonian world resembles the Einsteinian in rough outline. It is only after one has adopted the relativistic standpoint that it becomes apparent how, by making various simplifying assumptions, Einstein's formulas can be made to yield numerical results similar to Newton's. There is no transition from the system of Newton to the system of Einstein. One does not proceed from the first to the second by amassing data, perfecting measurements, and making slight adjustments to first principles. What is needed is some totally new ingredient. It is a "transcendental induction" and not an "amplifying induction" that leads the way from classical to relativistic physics. After the step has been taken, it is of course possible to recover Newtonian mechanics by reduction. Newton's astronomy can thus be seen to be a special case of Einstein's "pan-astronomy," just as Euclid's geometry is a special case of Lobachevski's pan-geometry.

II

As is well known, however, relativity did not come into its own as a general astronomy or cosmology. It grew out of Einstein's reflections upon the fundamental concepts of physics, his questioning of obvious ideas, his complexification of what appeared to be simple. What could be more immediate, obvious, or simple than the idea of simultaneity? Imagine a train consisting of a number of cars standing on a track consisting of two parallel rails. The engine starts, and all the cars seem to start moving at the same time. The example actually illustrates two of the fundamental concepts of prerelativistic physics, parallelism and simultaneity. We saw in the preceding chapter how Lobachevski's geometry challenged the intuitive notion of parallelism. Similarly, relativity challenges the intuitive notion of simultaneity. It does this by insisting that before we assert the simultaneity of events occuring at two different locations in space, we specify how we have determined that they are simultaneous. From this novel requirement relativity was born.

Relativity thus lays down a very simple challenge: If simultaneity is indeed a "clear and simple" idea, how is it used? How does one demonstrate simultaneity? How does one recognize it? How do observers in different frames of reference reach agreement about what events are simultaneous? In other words, how is the concept put to work? In which experimental judgments is it involved? For isn't the involvement of concepts in judgment precisely what we mean when we use the word *empirical?* And then, when we have answered these questions, when we have imagined a system of optical signals capable of allowing two different observers to agree on the simultaneity of two events, the relativist requires us to incorporate our experiment into our conceptualization. He reminds us that our conceptualization is an experiment. The world, then, is not so much our representation as our verification. Henceforth, a discursive and experimental account of simultaneity must be attached to the supposed intuition that originally revealed to us the coincidence of two phenomena. The primitive quality of the pure idea does not survive; we know it only through its composites. What seemed to be a primary idea has no basis in reason or experience. As Brunschvicg remarks, "it can neither be defined logically by self-sufficient reason nor observed physically in a positive form. It is in its essence a negation. It is tantamount to denying that a certain time is required for the propagation of a signal. Thus we see that the notion of absolute time, or more precisely the notion of a unique measure of time, i.e., of simultaneity independent of the frame of reference, owes its apparent simplicity and immediacy to a faulty analysis.[1]

The same critical principle underlies recent work of Werner Heisenberg (b. 1901). He, too, insists upon operational definition of such simple notions as the location of a point in space: It is illegitimate, says Heisenberg, to speak of the location of an electron if we cannot propose an experimen-

[1] Léon Brunschvicg, *L'expérience humaine et la causalité physique* (Paris: F. Alcan, 1922), p. 408.

tal method of locating it. In vain, realists will answer that we will find the electron where it is, trusting to the immediate, clear, and simple idea of location. But Heisenbergians will retort that it is a delicate experimental task to locate a minuscule object, and that the more precise the experiment is, the more it changes the location of the object. Experimentation is thus intimately involved in the definition of what is. Any definition is an experiment; any definition of a concept is functional. For both Heisenberg and Einstein the issue is one of finding experimental correlates of our rational ideas. These ideas therefore cease to be absolute, since they now stand in one-to-one correspondence with experiments of greater or lesser precision.

III

Thus, even notions whose essence is geometric, such as position and simultaneity, cannot be grasped in any simple way but only in composite, through experimentation. The new standards of rigor require us to go back to the experimental origins of geometry. Physics becomes a geometrical science and geometry a physical science. Naturally, notions more deeply enmeshed in material reality, such as mass, will take on a more composite form in the theory of relativity; there will be more than one kind of mass, for example. In this respect there is a very marked contrast between the new spirit and the old. I shall try next to explain the philosophical meaning of this.

In the science of centuries past, the unity of the notion of mass, its immediate and obvious character, derived from the vague intuition of quantity of matter. People had such confidence in the mind's concrete grasp of Nature that Newton's definitions seemed merely to give precision to a vague but well-founded idea. When Newton defined *mass* to be the quotient of a force divided by an acceleration (m = W/g — trans.), his formula was interpreted as indicating the specific role of the substance of the moving object: The more matter there was in an object, it was thought, the more effectively it opposed any force applied to it. Mass, in Newton's sense (or, in modern terms, inertial mass — trans.), was the measure of this opposition. Later, when Maupertuis defined *mass* as the quotient of impulse by velocity, we find the same vague thought, the same confused intuition, still exerting a powerful influence. Once again, the greater the quantity of matter in a particle, the more effectively it opposed an impulse applied to it. More theoretically, dimensional analysis suggested that the mass in both cases was the same, indicating the same coefficient of resistance, and no thought of a possible difference ever arose. Thus the primary notion of mass, well grounded in both theory and experiment, appears to have escaped all analysis. This simple idea seemed to correspond to a simple *nature*. On this point science apparently did no more than translate reality directly.

Dimensional analysis is not always as sovereign a determinant as is sometimes thought, however. Moreover, the claim to have an immediate grasp on the concrete is in many cases quite presumptuous. In the case in point, relativity has proved to be both less realist and more fertile than the older physics. Relativity complicated a simple notion by giving a mathematical structure to what had been concrete: The special theory proves that the mass of a moving object is a function of its velocity. But this function is not the same for mass in Maupertuis's sense as for mass in Newton's sense. The identity of the two masses holds true only to a first approximation. The two notions are identical only if one ignores their subtle notional structure. Dimensional analysis cannot discriminate between functions homogeneous with respect to velocity, and this is precisely the case for the correction coefficients, in which the velocity of the moving mass figures only in a quotient, divided by the velocity of light.

Relativity also contributed to a further elaboration of the Newtonian definition of mass. It introduced a distinction between the so-called longitudinal mass, or mass calculated along the particle's trajectory, and the transversal mass, calculated along a normal (i.e., perpendicular) to the trajectory (which is a kind of coefficient of resistance to distortion of the trajectory). A possible objection to this reasoning is that this distinction is artificial, since it is a consequence of an arbitrary decomposition of a vector. What is instructive to observe, however, is that this artifice, this decomposition, is even possible. This shows how far removed the new physics is from the old classical mechanics, in which mass, regarded as a fundamental unit, was by definition a simple entity.

In this special case it would again be quite easy to recover the classical mass as a special case of the relativistic masses. One has only to do away with the inner mathematics, the theoretical subtleties, that make relativity the complex form of rationalism that it is. To do so would bring us right back to the old simplified reality and simplistic rationalism. Newtonian mechanics can be deduced from Einsteinian mechanics by elimination, but the deductions cannot be reversed, either in detail or in the large.

If we look at specific concepts in order to determine just what was known in nineteenth-century physics and what is known now, it is striking that the concepts of modern physics are not only more precise but also more general than the concepts of the older physics, and that nothing is *simple* anymore unless we are willing to settle for *simplifications*. It used to be believed that concepts were simple in their pure form and became complicated when applied, whether cleverly or not so cleverly. In modern science the time for precision is not just the time of application; precision is required from the outset, from the first definition of principles and concepts. As Federigo Enriques rightly observes, modern "physics does not offer a verification of classical mechanics with enhanced precision but rather leads to a correction of the principles on which classical mechanics was based."[2] Here we see a reversal of the epistemological perspective, further examples of which will be adduced in what follows.

IV

The need for complexity is not always clear, and one might venture to predict how still simple concepts may become more complicated in the future. In this way we may gain some direct appreciation of the psychological anxiety that accompanies doubt about the fundamental concepts of a science. A good candidate for this exercise is the concept of velocity. Velocity (or its scalar correlate, speed) has emerged

[2] Federigo Enriques, *Les concepts fondamentaux de la science*, trans. L. Rougier (Paris: Flammarion, 1919), p. 267.

from relativistic manipulation more or less unscathed, though it has not been possible to justify the fact of a maximal speed (assumed in special relativity to be the speed of light in a vacuum — trans.). Formerly, when knowledge was compartmentalized into conceptual knowledge and applied knowledge, *a priori* principles and *a posteriori* principles, it was scarcely possible to admit that there could be a limit to the application of the concept of velocity. Now, non-Newtonian doctrine obliges us to regard a *fact,* the speed of light (in a vacuum), as an upper limit inscribed in the very principles of mechanics. If the speed of a moving object were to reach the speed of light, the mass of that object would be infinite. The absurdity of the conclusion entails the absurdity of the hypothesis. In a science whose concepts are mathematized, empirical notions fit together (*se solidarisent*) rationally. Such "interference" between notions of optics and mechanics may surprise the philosopher who believes that our intelligence finds its definitive structure in contact with a geometrical and mechanical world. Perhaps his surprise will diminish after he has read the next few chapters, in which I shall explain the construction of an *optical* intelligence, for which the facts of optical experience are formative.

In certain respects, however, the concept of velocity is more confused than it appears. What velocity actually means has gradually become less and less explicit as velocity has increasingly come to be identified with the kinematic notion of momentum. It is no longer possible to specify the mass of a moving object without taking its velocity into account; hence the notion of velocity has tended to be subsumed in the notion of an associated mass. Momentum in the kinematic sense is only an especially vivid special case of another momentum whose essence is more algebraic in nature. In view of these many difficulties, Niels Bohr recently stated that everything associated with the notion of velocity is shrouded in obscurity. Velocity is no longer a "clear and distinct idea" except for those who trust only in common sense.

One particular point that remains obscure is the meaning, in realistic terms, of velocity. Clearly something is moving, but one is no longer sure quite what it is. Consider, for example, the penetrating book by Karl Darrow entitled *The Synthesis of Waves and Particles.* We shall see shortly that when we study the speed of sound, which seems so clear when defined in textbooks, we are actually studying an ill-defined phenomenon. The same is true of the speed of light. It should come as less of a surprise, therefore, that when we study the dualistic phenomenon of matter waves and particles, we find ourselves faced with two different speeds. This, according to Darrow, leads to the proposition that "a flow of free electrons has two different velocities: one when we regard it as a set of particles, the other when we look upon it as a train of waves. Mustn't it be the case, however, that one of these two velocities is the right one, and isn't it possible to tell which one it is by measuring the actual time that the electrical current takes to cover a given distance? Let us examine this possibility. When we do, we shall discover that it isn't so easy after all to avoid such an ambiguity."[3] Thus in trying to assign a meaning to veloc-

[3] Karl K. Darrow, *La synthése des ondes et des corpuscules,* trans. Boll (1931), p. 22.

ity, we run up against the problem discussed in the introduction: that ambiguity attaches not to our knowledge of reality but to reality itself.

Isn't it also striking that one of the most serious errors in Aristotelian mechanics has to do with confusion over the role of velocity in motion? Aristotelianism in some sense ascribed too much reality to velocity with its proposition that a constant force was necessary to maintain a constant speed. As is well known, it was by limiting the role of velocity that Galileo founded modern mechanics. The first principle of relativity was established by assigning a theoretical role to the speed of light. In a still more recent development, matrix mechanics (one of the precursors of modern quantum theory — trans.), the formal role of momenta will, when further investigated, no doubt reveal additional derived senses that must be ascribed to the concept of velocity once regarded as primitive.

My purpose in reminding the reader of these revolutions in the interpretation of a single concept is to call attention to the fact that they coincide with more general revolutions that have profoundly affected the history of science. Concepts and conceptualization go hand in hand. We are dealing neither with words whose meanings change while the syntax of the language remains the same, nor with a free and changing syntax applied to the organization of unchanging ideas. Theoretical relations among notions modify the definition of those notions as much as changes in the definition of the notions alter their mutual relations. More philosophically, it may be asserted that thought changes in form when it changes in object. No doubt there are some kinds of knowledge that appear to be immutable. This leads some people to think that the stability of the contents is due to stability of the container, or, in other words, that the forms of rationality are permanent and no new method of rational thought is possible. But structure does not come from accumulation alone; the mass of immutable knowledge does not have as much functional importance as is sometimes assumed. If it is granted that scientific thought is, in its essence, a process of objectification, then it follows that its real motive force is rectification and extension. This is where the dynamic history of thought is written. *It is when a concept changes its meaning that it is most meaningful.* For it is then that it becomes, in all truth, an event, a conceptualization. From the didactic standpoint (the psychological importance of which is too often neglected), it is true to say that a student will better appreciate the point of Galileo's idea of velocity if the teacher has explained the role of velocity in Aristotle's theory of motion. For then he will appreciate the *psychological advance* that Galileo made. A similar statement may be made about the conceptual changes required by the theory of relativity. Non-Newtonian physics both subsumes classical mechanics and stands apart from it. It is not merely that the internal structure of the theory of relativity bestows upon it a static conceptual clarity. It is also that relativity throws a new and strange light on concepts that had once seemed not just clear but self-evident. The conviction that relativity calls forth is more powerful than the naive belief in reason's early successes, for the theory of relativity makes clear in what ways it marks a conceptual advance over the old physics. It demonstrates the superiority of the mature over the elementary. After relativity the scientific spirit is able to sit in judgment on its own past.

V

What makes people believe that the scientific spirit, despite the profoundest of changes, always remains fundamentally the same? The answer, I believe, is that the true role of mathematics in scientific thought has not been generally appreciated. It has been repeated endlessly that mathematics is a language, a mere means of expression. People have grown used to the idea that mathematics is a tool wielded by a self-conscious mind, mistress of a set of ideas endowed with premathematical clarity. Such a division may have made sense early in the history of science, when the images of intuition carried suggestive force that aided in the formulation of scientific theory. If one grants, for example, that the idea of attraction is an idea that is "clear and simple," it is reasonable to say that the mathematical formulation of the laws of attraction serves simply to specify certain instances of the general idea and to link together various of its consequences; the law of surfaces is much the same, it, too, being an expression of something whose intuitive meaning is straightforward and clear. The new science shuns naive images, however, and has in a sense become more homogeneous: It stems entirely from mathematics. Or, rather, it is mathematics that sets the pattern of discovery. Without mathematics science could not even conceptualize the phenomena of relativistic physics. As the noted physicist Paul Langevin (1872–1946) put it some years ago, "tensor calculus knows physics better than the physicist does." Psychologically, tensor calculus is the matrix of relativistic thinking. Contemporary physical science has been created by this mathematical instrument, much as microbiology was created by the microscope. None of the new knowledge is accessible to anyone who has not mastered the use of this new instrument.

Confronted with such a complex body of mathematics, the philosopher may find himself tempted to repeat the familiar charge of formalism. And indeed, when a mathematical law is discovered, it is rather easy to reformulate it in any number of ways. Intelligence then acquires such agility that it seems to go soaring over the realities, moving in a rarefied atmosphere of formal mathematical manipulation. But mathematical physics is not as independent of its object as doctrinaire believers in axiomatics maintain. To see this, we need only examine formalism from a psychological standpoint. Psychologically, a formal idea is an incomplete simplification, a limit-thought that is never actually attained (in the sense that a mathematical function may never attain its limit). In fact, any formal theory always has its substantive basis in tacit examples or veiled images. The mathematician tries to convince himself that the substantive example actually plays no deductive role. He proves this, however, only by showing that one example may be substituted for another. This is formally satisfactory but psychologically inadequate: The mind has a hard time grasping thoughts in a void. Protestations to the contrary notwithstanding, the algebraist always thinks more than he actually writes down. *A fortiori,* the mathematics of the new physics feeds on its empirical applications. It is certain that the psychological weight of Riemannian geometry increased when that geometry found application in the theory of relativity. Just as Newton's physics was perfectly attuned to Euclidean geometry, Einstein's physics appears to be perfectly attuned to Riemannian geometry.

By taking a systematic psychological approach we can hardly miss the way in which the mathematical tool affects the craftsman who uses it. *Homo mathematicus* is taking the place of *homo faber*. Tensor calculus, for example, is a marvelously flexible tool, and with it the mind acquires new capacity for generalization. Prior to the mathematical age, during the age of the solid, it was essential that reality offer the physicist an abundance of examples pinpointing the idea to be generalized: An idea was then a summary of experiments already carried out. In the new relativistic science, a single mathematical symbol rich with significance indicates the thousand traits of a hidden reality: An idea is now a program of experiments still to be carried out.

Besides the inductive and inventive power that tensor calculus bestows upon the mind, we must also consider, in order to complete our psychological characterization of this new mathematical tool, its value as an agent of synthesis. The tensor calculus in effect demands that we forget nothing, that we keep all variations of the symbol constantly in mind. This is a rational extension of the Cartesian use of enumeration as a mnemonic technique. I shall return to this point in the conclusion of the present work, where I intend to show how non-Newtonian science may be generalized to yield a non-Cartesian epistemology.

Awareness of the larger problem informs the very details of calculation, perpetuating the universal ideal enshrined in the principle of relativity. With relativity we have come a long way from the analytical state of mind characteristic of Newton's thought. It is in the realm of aesthetics that we may find synthetic values comparable to the symbols of mathematics. When we think of the beautiful symbols of mathematics, wherein the possible and the real are joined, the images that come to mind are Mallarméan:[4] "Their breath of inspiration and virgin accent! One dreams of them as of something that might have been. With reason, because none of the possibilities that hover about a figure, as ideas, must be neglected. They belong to the original, even against probability."[5] In the same way, purely mathematical possibilities belong to the real phenomenon, even against the first lessons of immediate experience. What might be, in the judgment of the mathematician, can always be tested by the physicist. What is possible and what is (*l'Être*) are homogeneous.

Wave and quantum mechanics have considerably accentuated the synthetic value of mathematical physics. Mathematically, these new disciplines are in many ways systematic procedures for achieving generality. The extreme generality of the Schrödinger equation is obvious at a glance.[6] The same is true of matrix mechanics. A pragmatic physicist — if any still exist — might raise a thousand objections against all the phantom terms[7] that appear only briefly in the course

[4] Stéphane Mallarmé (1842–1898), French symbolist poet.

[5] Stéphane Mallarmé, *Divagations*, p. 90.

[6] Named for Erwin Schrödinger (1887–1961), German physicist. Its solution is the so-called wave function of the physical system it represents. — Trans.

[7] Bachelard here refers, I think, to various mathematical "tricks" used to solve partial differential equations such as the Schrödinger equation. — Trans.

of calculation: Though formally necessary, they disappear without trace in the end. But it is a serious mistake to believe that these phantom terms are devoid of psychological reality! They are quite simply the indispensable props of thought. Without them scientific thought would seem to be nothing more than a bare compilation of empirical results. It is frequently these phantom terms that establish the idealist link (between ideas and things — trans.) and permit the replacement of causality by consequence — yet another important feature of the rationality of contemporary science.

Thus the scientific spirit cannot limit itself to conceptualizing the salient points of one particular experiment; rather, it must attempt to conceptualize all *possible* experiments. The point is a subtle one and difficult to make precise. Heisenberg has of course laid down the positivistic criterion that all theoretical notions must have experimental meaning. But if we look closely at what he says, we discover that he is willing to allow imaginary experiments. It is enough that the experiments in question be possible. Thus ultimately mathematical physics expresses itself in terms of experimental possibilities. The possible has in a sense drawn nearer to the real; it has recaptured a place and a role in the organization of experience. At the same time it has distanced itself from the rather fantastic constructions of the "philosophy of *as if*" (i.e., of reasoning based on analogy — trans.). With a mathematical organization of experimental possibilities in hand, it is but a short step back to the empirical. The real turns out to be a special case of the possible. The recent broadening of the horizons of science is probably most easily appreciated here.

To sum up, if we compare the general epistemological

frameworks of contemporary physics and Newtonian physics, it is clear that the old doctrines did not *develop* into the new, but rather that the new *enveloped* the old. Intellectual generations are nested, one within another. When we go from non-Newtonian physics to Newtonian physics, we do not encounter contradiction but we do experience contraction. It is this contraction that enables us to locate the limited phenomenon within the enveloping noumenon, the special case in the general; yet the particular can never evoke the general. Henceforth, to study phenomena one must engage in purely noumenal activity; it is mathematics that opens new avenues to experience.

Chapter Three

Matter and Radiation

I

As Whitehead justly remarks, "the terminology of physics is derived from seventeenth-century materialist ideas."[1] In my view, however, it would be a serious philosophical error to believe that materialism was genuinely concrete, especially when it presents itself, as it did in the seventeenth and eighteenth centuries, as an incompletely elaborated scientific theory purporting to be an immediate reflection of reality.

[1] Alfred North Whitehead, *Science and the Modern World*. The citation appears in the French translation by Ivery and Hollard on p. 200.

In fact, materialism is the product of an initial abstraction, which seemingly inevitably maims our notion of matter forever after. This abstraction, about which neither Baconian empiricism nor Cartesian dualism seems to harbor the slightest doubt, is that the spatial position of a material object can be exactly determined. Materialism also tends to limit the notion of matter in still another sense, by precluding so-called action at a distance, or in other words by refusing to allow a material object to produce effects at points in space other than where it is located. Thus materialism shades by degrees into realist atomism. Descartes's protestations to the contrary are scarcely convincing if matter alone has the attributes of being extended, composed of solids, associated with strictly local properties, and defined by a form, indeed inextricably bound up with a form. To correct for this utterly abstract and geometric localization, materialism equipped itself with a physics of fluids, exhalations, and spirits, but never questioned its original intuition. It was all too easy to ascribe movement to one vague fluid or another, whose sole function was to convey *elsewhere* the usual properties of matter.

Localizing matter in this way creates an artificial distinction between geometric and temporal properties. Phenomenological investigation then falls into one of two disciplines: either geometry or mechanics. Modern philosophers of science have understood the danger of such an arbitrary division. Schlick puts it quite well. It is impossible, he says, to speak "of a determinate geometry of space without taking account of physics and the behavior of natural bodies."[2] The

[2] Moritz Schlick, *Espace et temps dans la Physique contemporaine*, trans. Solovine, p. 33.

problem of the structure of matter must not be separated from the problem of its temporal behavior. It is apparent that the most obscure of metaphysical riddles attaches to the area where spatial and temporal properties intersect. What is the nature of this riddle? It is difficult to say, precisely because our language is materialistic, because we think for instance that any substance derives its fundamental nature from its material base, which remains unperturbed and indifferent to the passage of time. The language of space-time is no doubt more appropriate to the study of the classical *nature-law* (i.e., *physis-nomos* — trans.) synthesis, but this language has yet to develop a plentiful enough supply of imagery to have attracted the philosophers.

It should therefore be of considerable philosophical interest to study the scientists' efforts to work out new syntheses. The main concern of contemporary physics has in fact been to achieve a new and truly phenomenalist synthesis of matter and its actions. The physicists, in their attempts to link matter and radiation, have taught the metaphysicians a lesson in construction. We shall see shortly how open-minded physicists have been in their approach to radiation, even as they refused to accept that corrupt materialism which relies on theories of fluid motion, emanation, exhalation, and volatile spirits.

Let me state the problem in its most polemical, that is, metaphysical, form. Wurtz justifies atomism by invoking the ancient argument that it is impossible to "imagine movement without *something* that moves." To this argument microphysics might well retort that it is impossible to imagine a thing without positing *some action* of that thing.

And indeed there are crude and lazy forms of empiricism for which the objects of study are nothing more than inert bodies and for which experiments are not tests of theory but mere abstract observations, protestations of "concreteness" notwithstanding. Microphysics treats experimentation differently. It does not allow a distinction to be made between what is real now and what will be real at some time in the future. Only things in action can be described. What is a static photon, for instance? It is impossible to separate a photon from its ray, as "thing-oriented" (*chosiste*) observers, used to dealing with objects that can be handled physically, would no doubt like to do. The photon is plainly a thing-in-motion. More generally, it seems that the smaller an object is, the more fully it embodies the space-time complex, which is the very essence of the phenomenon. Broadening materialism so as to free it from its original geometric abstraction thus leads naturally to matter and radiation.

What, then, are the most important phenomenal characteristics of matter? Those related to energy. Matter should be seen, first of all, as a source and transformer of energy. Only then is it reasonable to ask how energy can take on the different characteristics of matter. In other words, the notion of energy is the most fruitful link between motion and objects. Energy is a measure of the efficacy of a moving object, and it is energy that tells us how *motion becomes an object*.

It is true, of course, that nineteenth-century physicists had already looked carefully at how energy is transformed, but they studied such phenomena in the large and not in microscopic detail (at least until the advent of statistical mechanics — trans.). Hence they concluded that the transformation of energy is continuous and that time has no structure. The situation was analogous to trying to understand a barter economy by studying savings banks. The problem of energy transfer was settled abstractly: In terms of our economic analogy, this solution was rather like explaining barter in terms of bills of exchange. But physicists believed that they had achieved an adequate account of the energy economy. Kinetic energy could be "stored" in the form of potential energy. Different forms of energy such as heat, light, and chemical, electrical, or mechanical energy could be converted back and forth in amounts measured by conversion coefficients. Physicists were of course aware that matter must somehow provide the locus or basis for this exchange of energy. But matter was often merely an occasional cause of some sort, a means of expression for a science whose aim was to remain faithful to a realist philosophy. What is more, a whole school (of thermodynamics — trans.) claimed to be able to do without the idea of matter. This was the time when Ostwald maintained that striking someone with a stick proves nothing about the existence of an external world.[3] The stick doesn't exist; only its kinetic energy does. Karl Pearson said much the same thing:[4] "Matter is non-matter in motion."[5] Assertions such as these seemed quite legitimate, since matter was seen merely as the unchanging support of phenomena involving an external quality, energy, which was assumed to be indifferent to the nature of its support, so that it was easy to make a Berkeley-style critique of materialism by arguing that the

[3] Wilhelm Ostwald (1853–1932), German physicist.
[4] Karl Pearson (1857–1936), English scientist and statistician.
[5] Cited by Reiser, "Mathematics and Emergent Evolution," *Monist*, October 1930, p. 523.

support was unnecessary and that what was truly essential in all thermodynamic phenomena was energy. Proponents of this "energist" view took little interest in investigating the "structure" of energy. They not only opposed the research of atomists into the structure of matter but in their own area favored the study of energy in general as opposed to attempts to "construct" energy from more elementary notions.

Brunschvicg has made some penetrating remarks about the parallelism that exists between the doctrines conservation of energy and conservation of mass: "Chemical substantialism, which takes its lead from the materialistic ontology of ancient atomism, seems to require a physical substantialism, which behind the diversity of qualitative appearances posits, as did the Stoics of old, the unity of causal reality. . . . A now commonplace idea is that . . . a sort of *causal substratum* underlies any transformation of a physical order; this is analogous to the strictly *material* substratum which, since Lavoisier's work in chemistry, has customarily been regarded as imperishable and eternal, beyond the compositions and decompositions of different bodies."[6] Thus both the realism of energy and the realism of matter were put forward in the nineteenth century as general philosophical doctrines. The effect of this was to push physics toward abstraction, continuing what has been called the "depopulation" of space and time, in contrast to modern doctrines whose "spatializing" and "enumerating" tendencies Brunschvicg has so clearly explained.[7]

[6] Brunschvicg, *L'expérience humaine et la causalité physique*, pp. 351–352.
[7] This is an apparent reference to statistical mechanics, which relies heavily on the idea of counting the molecular "population" of a region of phase "space." — Trans.

For the old physics of intuition, then, both matter and energy lack structure. The problem with this view is that it neglects the essential temporal aspect of energy. We cannot really understand energy until we understand phenomena that are extended in time. To say merely that matter has certain energy-related properties, that it can absorb, emit, or store energy, ultimately leads to contradictions. When energy is stored it becomes latent, potential, fictive, as when a sum of money is passed through the teller's window of a bank: This stored energy, which has real significance only when it reemerges after a certain time, thus becomes atemporal.

Modern physics has reunited energy and matter, bound them together in a sort of perpetual structural exchange. Contrast this with the vague, intuitive idea of energy being stored without substantial change to the storage medium: for example, heat energy stored in a lead bullet raising its temperature from zero to one hundred degrees, or chemical energy (from gunpowder) being converted to kinetic energy by firing a gun and accelerating the projectile to a speed of five hundred meters per second. Instead of this simple addition of energy to an unaltered object, modern physics imagines an ontological dialectic of matter and energy. The atom not only atomizes all atomic-scale phenomena but also structures whatever energy it emits. It is subject to discontinuous transformation stemming from discontinuous absorption or emission of energy. Hence we cannot simply say that matter is known to us through energy as substance is known to us through phenomena. Nor should we say that matter *has* energy, but rather that matter *is* energy and, conversely, energy *is* matter. The replacement of the verb *to have* by the verb *to be* is common in the new science. The metaphysical implications of this change are, I think, incalculable. It amounts to substituting equations for descriptions and quantities for qualities; yet the shift from qualitative to quantitative reasoning is in no sense a jettisoning of philosophy. On the contrary, it marks a decisive advance for mathematics, which recaptures a good deal of metaphysical terrain. The point is that a quantitative organization of reality has *more,* not less, content than a qualitative description of experience. Qualitative description is generally vague and inconsistent and invariably a one-sided summary. Focusing on quantitative fluctuations makes it possible to define what had been undefinable about particular qualities. In any case, naive qualitative realism has been shown repeatedly to be inadequate. For example, ionization studies have explained why the sky is blue and shifted the explanatory burden from matter to radiation. One possible objection is as follows: that the property one conceives of now as a property of radiation used to be conceived of as a property of matter — as when people used to say that a great enough thickness of air *is* blue. But this objection really misses the point. It is clear that the substantive ties have been loosened, and that nothing but language remains to tie us to immediate realism. The great vault of the heavens may seem azure to us, but we no longer regard that azure as a true substantial property. The blue of the sky has no more *existence* than the vault of the heavens.

The very fact that energy changes matter results in a peculiar shift of scientific language from metaphor to abstraction: It is because the atom absorbs or emits energy that it changes form, and not because it changes form that it absorbs or emits energy. If this subtle point seems hard to grasp, the reason is that we are apt to ascribe too great a causal role to individual atoms. This prevents us from conceptualizing the problem from the outset in probabilistic terms. In dealing with atomic and subatomic phenomena, we must try hard to shed our realist presuppositions; only then can we understand how an abstraction like a change in energy level can have explanatory value.

Thus the study of "microenergetics" seems to entail a *dematerialization of materialism*. Later on, we shall see how to talk about abstract configurations, configurations without figures, so to speak. The human imagination, first schooled by the study of spatial forms, later embraced the hyper-geometry of space-time. When we have absorbed this lesson we will be ready to look at why scientists are now trying to dispense with space-time itself and shift attention to the abstract theory of groups. With group theory we reach the ultimate abstraction, the realm in which relation has priority over being.

To sum up, then, the relationship between matter and energy is a good example of how the ontological value of scientific ideas is enhanced through interaction. It also shows how physicists have overcome the limitations of earlier intuitive views that placed too much stress on the spatial dimension and too much confidence in naive realism. Unlike matter, which to the naive observer seems localized, compact, and confined within a definite volume, energy cannot be pictured. It has no obvious structure and needs quantum theory to give it one (through its association of energy with number — the so-called quantum numbers). In its potential form, energy can occupy a volume with no precise limits and yet manifest itself in specific places. It is a marvelous concept, occupying a numerical middle ground between the potential and the actual, between space and time. In its energetic unfolding the atom is *becoming* as much as it is *being, motion* as much as it is *object*. In simple terms, the atom is the element wherein space-time comes to be (*l'élément du devenir-être schématisé dans l'espace-temps*).

Epistemology swings regularly back and forth between realism and nonrealism. So regularly, in fact, that it is possible to predict that physics will soon demonstrate that energy can become matter, just as it has recently shown how matter becomes energy. Normally cautious experimentalists have indeed suggested in recent years that it may be possible to create atoms out of kinetic energy. In an address to a society of industrial chemists delivered in New York (and what greater guarantee of positivism could there be than the combination of chemistry, industry, and the United States of America?), Millikan hypothesized that cosmic rays originate in an atom-building process occurring in the interstellar void.[8] Destruction of atoms in the stars yields radiant energy that is converted back into matter, electrons, under the conditions of low density and temperature characteristic of the interstellar vacuum. The positive and negative particles thus created go to make up various kinds of atoms, of which Millikan takes helium, oxygen, and silicon as typical examples. Cosmic rays are the result of this "reconversion" of energy to matter.[9]

[8] Robert Andrews Millikan (1868–1953), American physicist who was the first to measure the charge of an electron, in the famous Millikan oil-drop experiment. — Trans.

[9] See the article by Millikan appearing in *Revue générale des sciences*, October 1930, p. 578.

Millikan is at pains to point out that this alternation between motion and matter, radiation and particle, serves as a corrective to nineteenth-century ideas concerning the "death" of the universe.

This ontological convertibility of radiation into matter stands in some sense as a complement to the matter-radiation exchanges described in Einstein's theory of the photoelectric effect. Einstein's equation shows the absorption and subsequent emission of radiant energy by matter. Absorption and emission are reversible; they are described by the same equation.[10] Einstein's theory allows for the emission of energy by matter but not for the complete disappearance of the matter in question.[11] Similarly, radiant energy can be absorbed in matter, but there has to be at least a germ of matter to begin with. Thus Einstein's realism has a materialistic basis. With Millikan's intuition, the transformation of the real is more complete: He envisions pure kinetic energy creating its own material substrate. This occurs in conditions of such emptiness, such absence of all things, that it seems reasonable to say that matter is created from radiation, that an object is created by a motion (i.e., by kinetic energy — trans.). Einstein's equation is thus more than

[10] Presumably a reference to the equation that relates the frequency of radiant energy to photon energy, $E=h\upsilon$, where E is energy, h is Planck's constant, and υ is frequency. — Trans.

[11] This assertion is rather ambiguous. By "Einstein's theory" Bachelard seems to mean the theory of the photoelectric effect. The special theory of relativity does of course allow for the "complete disappearance" of matter, which can be converted into energy according to the famous formula $E=mc^2$. At the time Bachelard was writing, of course, no atomic bomb had yet been exploded to bring this point home. — Trans.

an equation of transformation, it is an ontological equation.[12] It obliges us to ascribe *existence* to radiation as much as to particles, to motion as much as to matter.

II

If we pursue the question of matter-energy exchange by delving more deeply into microphysics (where the new scientific spirit is taking shape), we discover that our common-sense intuitions are highly misleading: Even the simplest ideas, such as collision, reaction, and reflection (of light or matter), are in need of revision. In other words, our simple ideas need to be complicated before they can explain the phenomena of microphysics.

Consider, for example, the reflection of light. The idea of reflection, which seems so clear to our macroscopic intuition, becomes confused when the problem is to understand how radiation is ''reflected'' from a particle. Studying the ''photon-electron interaction'' will help us to understand why Descartes's simple ideas, the products of an overly hasty fusion of traditional geometrical and empirical teaching, are epistemologically inadequate to describe the new science.

The usual physics-class experiments with light rays and mirrors are at first glance so simple, so clear, so distinct, and so geometric that they might be taken as a paradigm of *scientific behavior,* where I use the term *behavior* in much the same spirit as Janet, who has described what is distinctive about human intelligence as ''basket behavior'': The intel-

[12] Here, the equation Bachelard has in mind does indeed seem to be $E=mc^2$. Cf. n. 11. — Trans.

ligence of the child far outstrips that of the dog, says Janet, because the child will collect objects in a basket, whereas the dog will not.[13] In fact, the mirror paradigm is such a primitive instance of scientific thought that it is rather difficult to analyze psychologically. Beginning science students are frequently astonished by the teacher's insistence that the experiment demonstrates the *law* of reflection. To them it seems perfectly obvious that "the angle of incidence equals the angle of reflection." There is nothing problematic about the immediate phenomenon. Priestley states in his history of optics that the law of reflection was always known and always understood. The difficulty of teaching the subject here stems, as is so often the case, from the simplicity of the experiment. The data of this experiment are precisely of that *immediate* variety which the new science wants to reconstruct. And the issue is not without importance, because the reflection of light by a mirror is often taken as a paradigm for collisions in general. Intuitions of the most diverse kinds reinforce one another: Elastic collisions are interpreted in terms of reflection by applying an intuitive principle stemming from Kepler, who hoped that "all natural phenomena might be related to the principle of light." Conversely, the reflection of light is explained in terms of light corpuscles colliding with and rebounding from the surface of the mirror. The parallel between the two cases has even been used to prove that light corpuscles are indeed material objects. Cheyne, a commentator on Newton's work, makes this point explicitly. Light is a body or substance, he says, because it "can be reflected and caused to change its direc-

[13] Pierre Janet (1859-1947), French psychologist and neurologist.

tion like other bodies, and the laws of reflection are the same as for these other bodies." In Metzger's very scholarly study (from which I have borrowed this citation), one finds passages in which the substantiality of light corpuscles is accentuated even more.[14] The rebound image always underlies the argument. The principle of sufficient reason clearly plays a part in establishing the law of reflection. It cuts short the problem of relating actual experiment to mathematical law: A particular experiment, which is held to be both fully explicated and richly explanatory, is declared to be *privileged,* and a whole theory is made to rest upon it. An event in the physical world is promoted to the status of an instrument of thought, a *Denkmittel,* a category of the scientific spirit. That this event happens to involve a startling "geometrization" ought to arouse the suspicion of any philosopher familiar with the complexity of mathematical physics.

Unfortunately, the privileged intuition that seems to cast such a clear light on the phenomenon of reflection may actually induce blindness. I shall illustrate what I mean by considering how the mirror paradigm introduces impediments to solving the problem of why the sky is blue.

The problem was first stated in scientific terms by Tyndall,[15] who became impatient with the curiously ambiguous substantialist explanation, according to which a thin layer of air is colorless but a thick layer colored — a typical instance of prescientific lack of discomfort with the simultaneous assertion of two realistic but apparently contradic-

[14] Hélène Metzger, *Newton, Stahl, Boerhaave et la doctrine chimique,* pp. 74ff.
[15] John Tyndall (1820–1893), British physicist.

tory propositions. Based on ingenious experiments with suspensions of mastic in clear water, Tyndall thought he could prove that the blueness of the sky was the result of the scattering of light by material particles. In 1897 Lord Rayleigh[16] gave a theoretical explanation showing that the light was scattered not by dust particles or water droplets but by the very molecules of gas that make up the atmosphere. According to this theory, all the light emitted by the sun is in fact scattered, but since the intensity of the scattered light is inversely proportional to the fourth power of the wavelength, it is the blue light, whose wavelength is the smallest, that predominates overall.[17] Rayleigh's formula is ingenious and the result of elaborate research, but the basic intuition is still quite simple: All of the incident energy is transmitted; the molecules in the atmosphere act merely as obstacles to the passage of light, which they reflect in the manner of a mirror. There seems to be no need to press the investigation any further. The basic phenomenon is perfectly clear, distinct, and intuitively obvious: moving objects bouncing off *things*.

But a very important fact remained hidden by the explanation itself. It would seem obvious the change in the color of the scattered light ought to have suggested the need for a spectroscopic analysis. Yet for a long time this need was neglected. Although numerous experimentalists investigated the intensity and polarization of the scattered light in the Tyndall phenomenon, "it is quite remarkable," as

[16] John William Strutt, Third Baron Rayleigh (1842–1919), English mathematician and physicist.
[17] This is a loose and somewhat inaccurate statement of the scattering law, but there is no need to give the full formula here. — Trans.

Victor Henri rightly observes, "that none of the students of that phenomenon had the idea of setting up a spectrograph to analyze the nature of the scattered light. . . . It was not until 1928 that an ingenious Hindu physicist, Sir Raman, pointed out that the scattered light contained components of lower and higher frequency than the incident light."[18] The scientific impact of the discovery of the Raman effect is well known, but its metaphysical importance is also considerable. Microphysics explains what is going on by pointing to an interaction between the incident radiation and the molecules of the atmosphere. The molecules add their own characteristic radiation to the incident radiation. The vibration that impinges on the molecule is not reflected as from an inert object, still less as a muffled echo of some sort. Rather, its "timbre" will be changed, so to speak, as energy at many new frequencies is added to it. But even this account is still too materialistic to encompass the quantum interpretation of the phenomenon. What actually emerges from a molecule on which a beam of light impinges? Is it really a light spectrum? Or isn't it rather a *spectrum of numbers* that conveys to us the novel mathematics of a new world? Be that as it may, when one studies the new quantum methods in depth, it becomes clear that the problem is no longer one of collision, rebound, and reflection, nor is it a simple exchange of energy; rather, there is a complex set of interactions between light energy and molecular energy, governed by complicated numerical rules. The mathematical interpretation of the sky's color is currently

[18] Victor Henri, *Matière et énergie* (1933), p. 24. Sir Chandrasekhara Raman (1888–?), Indian physicist.

a theoretical question whose importance cannot be overestimated. The blueness of the heavens, to which, as I said earlier, it is scarcely possible to ascribe any "reality," is as instructive to the new scientific spirit as the starry heavens were to physicists centuries ago.

In summary, then, it is only when science begins to struggle against its first intuitions, when it begins to study light phenomena in more than a limited and schematic way and seeks occasions to carry out new and different kinds of experiments, that it yields theories capable of correcting erroneous hypotheses and experiments capable of rectifying misleading observations.

III

The same problem of essential complexity comes up when we interpret the Compton effect (the scattering of photons by electrons — trans.) in terms of wave mechanics. An encounter between a photon and an electron changes the frequency of both.[19] In other words, the fact that these two "geometrical objects" occupy the same space at the same time has consequences for their "temporal" properties. Such an encounter is therefore not a mechanical collision, nor is it an optical reflection interpretable in terms of the mirror paradigm. It is an event about which we still know all too little, generally referred to by a singularly inapt name: electromagnetic collision. It is a phenomenon that combines relativistic mechanics, optics, and electromagnetism and that

[19] Here, the "frequency" of an electron or photon is the quantity E/h where E is energy and h is Planck's constant. — Trans.

is best described in the language of space-time. What poet will furnish us with the metaphors for which this new language cries out? How can we possibly imagine the amalgamation of space and time? What supreme view of harmony will enable us to accord repetition in time with symmetry in space?

There are positive experiments that illustrate the effects of "rhythm" [or frequency: Bachelard throughout uses rhythm as a metaphor for frequency — trans.] on "structure." For example, there is no known chemical procedure for separating the two isotopes of chlorine. Take any compound of chlorine you like: Ordinary chemical manipulations always yield the same mix of chlorine 35 and chlorine 37. But if one takes a sample of phosgene ($COCl_2$) and exposes it to a beam of ultraviolet light whose frequency coincides with the resonance band of the chlorine 35 isotope, the molecules of phosgene containing that isotope dissociate to release just chlorine 35 and not chlorine 37. The chlorine 37 remains in compound, unaffected by light of the wrong frequency.[20] This example shows how radiation can cause matter to be liberated. We do not yet fully understand such frequency-dependent reactions, because our temporal intuitions are still inadequate, limited by the ideas of absolute beginning and continuous duration. Such an unstructured idea of time would seem at first glance to allow no discrimination between frequencies. But this simple view is illusory because it sees time as continuous, whereas in microphysics time

[20] See V. Henri and Nowell, *Proceedings of the Royal Society of London*, 1930, no. 128, p. 192. Cited by Henri, *Matière et énergie*, p. 235.

is discontinuous.[21] Time acts through repetition more than through duration. A moment's reflection is enough to convince us of the fact that in the selective decomposition of phosgene, the temporal complexity is of an entirely different order from, say, the action of light on an explosive mixture of chlorine and hydrogen (as it was understood in the nineteenth century). Light is an excellent "rhythmic agent" (i.e., light of a particular "color" has a specific frequency — trans.) for probing the space-time complex of matter. Jean Perrin in 1925 proposed the hypothesis that all chemical reactions are photochemical reactions. No structural change of a substance is possible, he said, except through the agency of some form of radiant energy, and therefore of quantized energy, energy of a definite frequency: as if structures can be altered only by "rhythms." Such a hypothesis denies explanatory value to the macroscopic idea of collision. Subsequently, Perrin himself has suggested that collision should be reinstated as a possible cause of reaction, but he still maintains that there is a kind of causal equivalence between the energy of the collision and the energy of the radiation.[22]

Such an equivalence, it seems to me, is likely to lead to profound changes in our realist concept of chemical substance. Once radiation is allowed to act as an intermediary between molecules and regarded as an integral part of the real, it becomes possible to distinguish between substances that would otherwise appear to be identical. A molecule that

[21] When Bachelard says "time is discontinuous," he is actually referring to the fact that frequency, that is, energy, is quantized. The more radical notion that time itself is quantized is not, as it seems to me, intended. — Trans.

[22] See Haissinsky, *L'atomistique moderne et la chimie*, p. 311.

has absorbed a quantum of radiant energy is different from one that has not. The chemist thus finds himself confronted with a complex ensemble of matter-cum-energy of which the only possible description is statistical, since the molecules are all different and the distribution of energy is not uniform. As kinetic chemistry develops, energy factors are gradually assuming greater and greater prominence. ''Microenergetics'' is in fact nothing other than statistical quantum mechanics. In this connection it makes sense to speak of a statistical ontology of substances.

IV

It is time now to take a broader overview of the subject. Starting with the orbital shell picture of atoms, I want to trace the course of the subtle transition from a realist to a probabilistic view.

Scientists gradually came to realize that the ordering of the elements in Mendeleev's periodic table[23] was related to the number of electrons associated with each element. Until the advent of the quantum theory this general explanation of the system of elements was realism's greatest triumph. The basis of the explanation was the *real presence* of electrons in the atom. Little by little it was realized that the *location* of the electrons was also important, and the structure of the various orbital shells provided the key to explaining the distribution of the elements in the periodic table. Thus, at this stage of the theory, structure was assigned a realistic role more important than that ascribed to

[23] Named for Dmitri Ivanovich Mendeleev (1834–1907), Russian chemist.

the fundamental particles themselves. This theory of orbital shells eventually became the basis of a whole theory of chemical valence, which provided a way of explaining the various affinities among elements and went some way toward explaining chemical reactions in general.

At this point the whole vast realist architecture came into contact with a complex and subtle mathematics. Instead of ascribing properties and forces directly to electrons, physicists assigned quantum numbers and from the distribution of these numbers deduced the (orbital) locations of the electrons within atoms and molecules. Notice, here, how realism suddenly evaporates: *Number becomes an attribute, a predicate, of substance.* Four quantum numbers are all that is needed to identify an individual electron. The mathematics respects this individuality, moreover. Here we discover what is in effect the social law of substantial association: In any given atom, no electron is allowed to take on exactly the same set of quantum numbers as any other electron. Two different electrons in the same atom must differ in at least one of the four quantum numbers. It is these numbers that determine the "role" of the electron in the atom. This is the philosophical meaning of the Pauli exclusion principle.[24] It is clear that this principle contradicts any attempt to argue that the quantum properties are substantial, deeply ingrained in the substance of the electrons themselves, for the numbers are in a sense attributes "in extension." The only thing to prevent a given electron from taking on a particular set of quantum numbers is the fact that *another* electron already occupies that position. Given that the trend in con-

[24] Named for Wolfgang Pauli (1900–1958), German physicist.

temporary chemistry has been to extend the Pauli principle not simply to molecules but to all material amalgams (see the work of Fermi, for example), it follows that the organization of matter is in a sense synonymous with the quantum principle of individuation of its constituent elements. Wherever there is true organization, there is reason to bring the Pauli principle into play. In philosophical terms, this principle entails systematic exclusion of the *same* and systematic appeal to the *other*. Within any system, or, better still, in order for a set of elements to constitute a system, the components must be mathematically diverse. Only nonreacting chemical substances can be identical, for these are like closed worlds, existing in total indifference to one another.

What, then, characterizes a simple or compound chemical substance? Nothing other than this subtle organization of (quantum) numbers, this pattern of numbers that complement one another by obeying the rule that there must be no duplication among them. Thus there occurs a surreptitious transition from the *corps chimique* or chemical substance to the *corps arithmétique* in the technical sense of the term (which unfortunately translates into English as "field of rational numbers," ruining the play on words — trans.). A chemical substance or *corps chimique* is therefore a *corpus* of laws, an enumeration of numerical characters. This marks the first step in the transition from materialistic realism to mathematical realism.

The attribution of four quantum numbers to each electron has to be desubstantialized still further, however. The crucial point is that this attribution is actually statistical in nature; the need for a statistical justification of the Pauli principle is fairly clearly understood. The issue at this point remains confused, however. What seems clear is that the quantum numbers serve to quantify energy. But all energy attributions now appear to have a statistical origin. When the time comes to analyze energy interactions between matter and radiation, probabilities will again enter into the picture. Thus quantum arithmetic turns step by step into a kind of statistical arithmetic.

Let us consider, then, a chemical substance in terms of the complex mathematical description provided by quantum mechanics. It now represents little more than a *likelihood of reaction.* If one were to insist upon an ultraprecise definition of the term *reaction,* understood in a fully detailed manner in terms of the various transfers of energy involved, the so-called chemical substance would vanish altogether, leaving behind nothing more substantial than a card player's hope of winning a hand of poker. There are of course factors tending to promote stability, but these have to be deduced from the law of large numbers. And there are of course solid empirical facts, but these must allow for a certain minimal uncertainty. One can be sure that chlorine mixed with hydrogen will react, and one can even study the effect of light on the kinetics of their reaction. But to hope for a fully detailed quantum description, a precise specification of the energy level of each electron in each atom over the course of the reaction, is no more realistic than to hope for an exact account of the distribution of cards through a long night of bridge (presumably this refers to a *predictive* account, for an *ex post* account of the card game would not be difficult to give — trans.). Ultimately, chemistry must measure its certainties using the calculus of probabilities.

Thus chemistry, which for a long time was the substantialist science *par excellence,* has had the substance slowly drained out of it. If objects are judged in terms of the evidence for their objectivity, it must be granted that the objects of chemistry have been increasingly mathematized, and that there has been a remarkable convergence of experimental proof and mathematical proof. The width of the metaphysical abyss between the mind and the outside world, so great that the metaphysics of immediate intuition could never bridge it, is, I think, less formidable for a discursive metaphysics that tries to follow the progress of science. It is even possible to conceive of a veritable displacement of the real, a purification of realism, a metaphysical sublimation of matter. The procedure is as follows: First transform reality into mathematical realism, then dissolve mathematical realism in the new statistical realism of quantum mechanics. The philosopher who follows the discipline of quantum theory — the *schola quantorum* — must be willing to conceptualize reality in terms of its mathematical organization; more than that, he must accustom himself to taking the metaphysical measure of the real in terms of the possible, which is precisely the reverse of what the realist philosopher does. Let me therefore sum up the supremacy of number over thing and of probability over number in the following polemical formula: A chemical substance is but the shadow of a number (*l'ombre d'un nombre*).

Chapter Four

Waves and Particles

It is perhaps in connection with wave-particle dualism that the psychological observations that are the main feature of this book appear to be most firmly established. For it is here that we become aware of how poorly schooled we are by immediate experience and how much we suffer from the one-sidedness of our initial exposure to mechanics. The early resistance to Louis de Broglie's brilliant insights[1] can be un-

[1] Louis Victor de Broglie (b. 1892), French physicist and discoverer of the wave nature of electrons.

derstood as the results of psychological resistance that kept people from seeing the point of what experience had to teach: first, that we have as much to learn from fluids as from solids, and second, that we must learn to conceptualize solids in terms of our prior empirical understanding of fluids, if only to compensate for the epistemological consequences of the traditional approach, which proceeds in precisely the opposite direction.

Heisenberg quite rightly puts his criticisms of the traditional approach to pedagogical use, showing why we need to think in terms of waves as well as particles. In his *Physical Principles of Quantum Theory* one finds two curiously antagonistic chapters. The first of these contains a critique of particle theory based on ideas drawn from wave theory, which is accorded a kind of *a priori* validity. But the next chapter turns this critique on its head. It contains a critique of wave theory based on ideas drawn from particle theory, it, too, now taken to be valid. If this double critique were realistic in essence, we would at this point be caught in a vicious circle.

In fact, however, this dialectical critique is an excellent lesson in phenomenalist philosophy. And such a critique is necessary if the real problems are to be correctly formulated, without any unwarranted residue of realism. Perusing Heisenberg's two chapters is a psychologically salutary exercise. The first sets before us the paradoxes of wave mechanics: Mechanics actually has to be constructed from optics. Such notions as velocity, particle, energy, and position have to be explained or constructed. They are no longer clear, simple, immediate, and distinct. They no longer have explanatory value, which has been transferred to the wave.

For example, "the fact that the position of an electron is known up to a certain error Δq is to be interpreted, from the standpoint of the wave theory, as a wave function whose amplitude is different from zero only in a small interval of length approximately equal to Δq. Such a wave function can always be thought of as composed of a sum of elementary wave functions that reinforce one another in the interval Δq and cancel one another everywhere else, as in the case of interference."[2] What this technique amounts to, then, is a way of constructing a particle from a "wave packet," much as the kinetic theory of gases constructs pressure out of a series of collisions. Philosophically, this is of course an inversion of the usual realist procedure, an inversion that is normally prohibited. The immediate, real object is described indirectly as a composite of other objects; the particle is said to be a complex entity, constructed by synthesis rather than isolated by analysis. The wave-mechanical critique implies that the particle is no more real than the wave packet that makes it perceptible. Temporal events (i.e., waves — trans.) are fundamental to its being. Nothing of absolute permanence attaches to the particle, which is distinguished in this respect from what philosophers call substance (substance being, presumably, the permanent substrate of all its attributes). The waves that go to make up any particle must satisfy boundary conditions established in regions of space far removed from the point at which the material particle appears as a kind of emphemeral shadow. In other words, the existence of the particle is rooted in all

[2] Werner Heisenberg, *Principes physiques de la théorie des quanta*, trans. B. Champion and E. Hochard (Paris: Gauthier-Villars, 1932).

space. Long ago, Leibniz said, "*Quod non agit, non existit.*" Now the time has come to restate this aphorism in a positive form: Wherever the point acts, it exists. As de Broglie says, in wave mechanics "one does not conceive of the material point as a static entity involving at most a very limited region of space but as the center of a periodic phenomenon spread all around it."[3]

The next question is how to assign a velocity to a particle, when that particle no longer has an enduring identity. One after another, all the images of traditional mechanics lose their clarity. A particle that cannot be identified or located has no trajectory; its very substance is not governed by the principle of identity, the most fundamental of all conservation principles. This new particle, which is nothing more than a sum of vibrations, is *reconstructed rather than conserved.* Finally, we must be careful not to attribute qualities directly to the particle but rather to the constructive process.

Thus the revolution of wave mechanics has made what was once direct indirect, mediated what was once immediate, and replaced the simple with the complex. Psychologically, the new physics forces us to unlearn what we have learned, to undo one intuition with another, and to dispense with our first analysis so that we may think instead in terms of our new composite view of what matter is.

A particle must not be thought of as a small sphere of definite volume. Since it is impossible, for example, to conceive of making measurements inside an electron, the interior of the electron is like a closed preserve. To be quite rigorous,

[3] Louis de Broglie, "La nouvelle dynamique des quanta," in *Electrons et photons* (1928), p. 105.

this prohibition should be incorporated into the axioms of mathematical physics. And this is precisely what Coppel, Fournier, and Yovanovitch have proposed.[4] They point out that the existence of such forbidden regions of space makes it impossible to establish a correspondence between matter-filled space and the arithmetic continuum. The Archimedean hypothesis is false, in other words. Stated in geometrical terms, this hypothesis is as follows: Given two line segments, it is always possible to find some multiple of the smaller segment whose length is larger than that of the larger segment — given enough inches, in other words, one can cover any distance. Clear and intuitive as this hypothesis is, it no longer makes sense if certain regions of space cannot be measured. Once we enter this forbidden territory, we must renounce the principles of continuous measurement. This suggests that possibility of a non-Archimedean geometry. Such a geometry would have the advantage of incorporating the hard-to-measure substance into the system of measurement: "Physics imposes upon the Archimedean world the extralogical notion of substance, whereas within the framework of a non-Archimedean geometry, substance itself is reduced to the fundamental logical notions of space and time"[5] In other words, substance is subsumed in the hiatus of measurement, but this hiatus is not irrational, because it has been incorporated into the rules of rational explanation. This is a good example of the kind of flexibility that has been introduced into rational theorizing by the various dialectics affecting our choice of axioms. By a judi-

[4] T. Coppel, G. Fournier, and D. K. Yovanovitch, *Quelques suggestions concernant la matière et le rayonnement* (Paris: A. Blanchard, 1928).
[5] Ibid., p. 23.

cious choice of axioms it may be possible to eliminate irrational elements altogether. Thus "the irrational" is not an absolute. The more unfettered the spirit, the less opaque irrationality becomes.

In fact, the ingenious suggestions of Coppel, Fournier, and Yovanovitch have not been fully developed. The forbidden interior of the particle is actually swallowed up by uncertainty about the particle's location. Non-Archimedean measures may find a role in the study of regions of space filled with static particles. When the particles are in motion, however, the interaction of matter and motion complicates matters. This brings us to Heisenberg's work on the limitations of physical measurements.

II

Let us turn now, therefore, to Heisenberg's second chapter, in which the particle theory, assumed to be correct, is used to criticize the wave theory. It is quite difficult to carry out such a critique in a fully modern fashion; with the particle theory perhaps more than any other, old psychological habits deprive us of the suppleness of mind necessary to keep in step with contemporary science. The idea of describing waves in terms of the motion of a medium composed of presumably real "material points" is as old as the wave theory of light. Ever since Huyghens physicists have tried to explain the propagation of light in terms of the vibrations of some sort of material medium. Many theories treated this medium (for mathematical purposes) as a continuum but actually conceived of it as being composed of discrete particles. Other theories, also numerous, held that the medium, or the ether as it was sometimes called, was actually discontinuous. Proponents of the various wave theories claimed to be studying the *continuous* propagation of light, but their intuitive understanding reflected awareness only of motions embodied in distinct, fixed particles or "corpuscles." Propagation of the light from one particle to another was expressed in terms of mathematical formulas based to one degree or another upon intuition. In short, the wave theory of the old physics was far from complete, despite the specious clarity of the answers it was able to provide.

Be that as it may, Heisenberg's criticism of the wave theory parallels his previous criticism of the particle theory. He points out that such characteristics of waves as amplitude, period, and phase "originate in everyday experiences such as the observation of waves in water or of the vibrations of elastic bodies."[6] Thus these characteristics exhibit themselves as properties not of particles but of complex, deformable substrates. If we believe intuitively that the world is actually composed of particles, then this substrate must seem to be a composite, rather than a simple, phenomenon. Wave theories were used to explain the propagation of light (more by inference than by observation) and, more precisely, to account for the results of experiments with diffraction and interference. Ultimately, similar ideas were successfully applied to recently discovered phenomena involving so-called matter waves. The epistemological question remains, however: Do these successes justify the construction?

The crucial question is the following: Can we transfer to

[6] Heisenberg, *Principes*, p. 39.

waves such as those associated with the names of Fresnel[7] and de Broglie — waves whose existence we infer — all the immediate phenomenological characteristics of ordinary waves, such as those caused by tossing a rock into a pond? This question is strictly analogous to one we raised earlier, when we asked whether an electron could truly be said to possess all the properties of a material particle. And the answer is also the same: Just as it was impossible to specify the exact position of an electron in space, so too is it impossible to know the exact amplitude of a wave at every point. Any experiment designed to measure this amplitude can at best provide an average value for a certain region of space and a certain interval of time, and these intervals can never be reduced to zero (that is, to a precise point and a precise instant). In other words, it is illegitimate to think of a wave as having a material support, as being somehow equivalent to the oscillatory motion of a real element of matter. The old physics erred, therefore, in ascribing oscillatory properties to material points. This is the philosophical explanation of the failure of the old physics to construct a discontinuous ether. For even the proponents of the discontinuous ether theory believed intuitively that the existence of waves implies an extended medium and a continuum of points. When the time came to translate this notion into a continuous probability function, proponents of the ether theory were bound by initial presuppositions derived from their synthetic notion of what a wave actually was.

What we have, then, is a pair of images — waves and particles — that do not fit together in any simple way. Each is

[7] Augustin Jean Fresnel (1788–1827), French physicist.

clear only when considered apart from the other. In short, both are condemned to remain mere images; neither can claim to represent the deep, underlying reality. Yet these images can be instructive if we know how to use them as a source of analogies, if we practice conceptualizing each in terms of the other, and if we use each to limit the other's application. Both have proven themselves: the particle theory in mechanics, the wave theory in physical optics.

For students of scientific psychology, mechanical intuition has long been the primary focus of interest. Now that the wave theory has come into its own, however, there is genuine pedagogical interest in this subject as well. Nothing points up the psychological importance of the question better than the following remarks of C. G. Darwin:

> Something quite different from simple fundamental principles is needed: in particular, we must develop new forms of thought that will enable us to study phenomena too complicated to be treated fully by means of mechanics. In forging these new forms of thought we must, I believe, take account of the fact that the human mind is possessed of a very great inertia and, I might add, of a very great viscosity. It is always very slow to move from one equilibrium position to another. . . . If we wish to achieve equilibrium more rapidly, we must for a very short time apply a force far greater than is strictly necessary to accomplish our end. That is why I believe that the best course of action for the present is to emphasize the wave aspect of the theory to the detriment of the dynamic aspect, hoping thereby to achieve, in the shortest possible time, a proper balance between the two.

Once this equilibrium is established, Darwin goes on to say, a curious discovery lies in store:

> For problems involving particles (or at any rate what we take to be particles) we are obliged to use the methods of the wave theory, whereas for light, which seems beyond any doubt to have the character of a wave, we are obliged to use the particle theory.[8]

The pedagogical problem has both positive and negative aspects. On the positive side we must emphasize all that can be learned from the wave theory. On the negative side we must work to undo the naive realism that students develop through their work with the elementary ballistics of projectiles. One might, for instance, try to bring out what is gratuitous and not fully thought out in the ascription of reality to light particles. Some physicists have been too quick to suggest that the idea of the photon has restored Newton's old idea of a light corpuscle. A "restoration" of this sort is possible in the early stages of development of scientific culture, when it seems perfectly possible to substitute one intuition for another; but "rectified" ideas, the products of a mature science, are never identical with the ideas from which they developed. Every attempt to demonstrate mechanical interactions *among photons* has ended in failure. It is true that physicists have been able to observe the electron-photon interaction in the Compton effect; but attempts to study the collision of two photons have yielded only negative results. The method was to arrange two light rays so that they intersected. Rare as the photons may be along each

[8] C. G. Darwin, "La théorie ondulatoire de la matière," *Annales de l'institut Henri-Poincaré*, fasc. I, vol. 1, pp. 25–26.

ray, there is no apparent reason why collisions should not occur at the point of intersection. Yet the fact remains that no photons are found in the region between the two rays. This negative result can be summed up in the following philosophical proposition: It is impossible to demonstrate a "mechanical composition" of light particles, whereas the phenomenon of interference clearly demonstrates the effects of "composing" light waves.

Let us next recall (again for negative pedagogical purposes) the mechanically anomalous nature of the photon. Its mass would be zero if only it were possible to imagine it at rest. Its natural velocity is the maximal velocity that no material body is allowed to attain: the speed of light. The location of a photon in a bundle of light rays is of course subject to the Heisenberg uncertainty principle.[9] The photon thus embodies some of the same arbitrary contradictions as the ether in various theories of that medium. For example, in certain realist theories it was found necessary to assume that the ether is extremely light but at the same time highly elastic: thinner than any gas but more elastic than steel. Materialistic theories of light have for centuries been afflicted by such apparent empirical contradictions. Taken together, these difficulties might suggest to a philosopher the idea that the photon cannot be fully comprehended within the terms of a particle theory. The *material realization* of the photon turns out to be a flawed intuition. On the other hand, when the time comes to ask the physicist for a detailed account of the *wave picture* of the electron, these remarks should

[9] The Heisenberg uncertainty principle says that we cannot know *both* the position and momentum of a particle with absolute precision. — Trans.

make us less exigent.

No matter whether we are dealing with photons, electrons, or atoms, we must convince ourselves that it is better in general to speak of "realization" than of "reality." As Margenau says, "recognition of the fact that the realistic appeal of certain natural facts depends in large part on our modes of comprehension deprives naive realism of much of its persuasive power."[10] Experimental realization depends in the first place on our modes of intellectual apprehension. The first steps must be taken by theory. The phenomena of microphysics lack what Margenau calls realistic appeal (in English in the original — trans.).

When we have learned to balance our intuitive knowledge of waves against our intuitive knowledge of particles and to overcome the appeal of naive realism, which wants to assign enduring characters to all things, and, further, when we have understood the power of experimentation informed by theory (*l'expérience réalisante*), then we are ready to undertake a more dispassionate consideration of the wave-particle dialectic. Why would anyone want to find a causal connection between waves and particles if the whole problem were simply a matter of two different images, two views of a complex phenomenon? The theory of a "pilot wave" guiding the particle is actually little more than a metaphor for establishing a connection between the wave and the particle. The most that can be said is that this connection is neither causal nor substantive. The particle and the wave are not linked by any sort of mechanism. The connection between them is of a mathematical order; each should be

[10] Margenau, *Monist*, July 1929.

understood as a different "moment" in the mathematization of experience.

The conflict is attenuated by recent theories that interpret the wave associated with a particle as representing the probability that the particle will be found at a given location. According to this interpretation the wave is really a wave function, a mathematical expression defined on a so-called phase space, which unlike ordinary intuitive space has more than three dimensions. There is a natural way of transferring conclusions about this artificial space back to ordinary space, which for the new physics is nothing more than an illustrative device, a convenient place to which images can be attached, but never an adequate canvas for displaying the full set of quantum relations. The existence of this phase space raises philosophical questions that provide an opportunity to revise our idea of reality. A common objection is that a mathematical object like phase space is merely artificial.[11] For mathematical purposes, however, such artificial objects offer a maximum of generality, homogeneity, and symmetry. From a synthetic point of view phase space is in a sense more real than ordinary space. Phase space may be regarded as a sort of *a priori* form for creating models. If one wants to model a system consisting of many particles, phase space is indispensable. And it is an almost natural choice for statistical analysis, for statistical methods of course require large numbers of elements and therefore a space of many dimensions. The probabilistic interpretation of the wave function

[11] The English physicist Sir James Jeans (1877–1946) rightly states that a ten-dimensional space is neither more nor less real than our usual three-dimensional space. See *The Mysterious Universe* (New York: Macmillan, 1931), p. 129.

is most naturally expressed in terms of phase space. That interpretation can then be recast in terms of ordinary space, filled with matter slow and heavy enough to ensure that steady-state laws are not affected by random fluctuations. The empirical data of macrophysics are insufficient to serve as a guide, however; overly realistic, these data have to be reworked before a statistical interpretation can be given. In the previous chapter, where we looked briefly at the mathematical ideas that are gradually reshaping chemical theory, I stated as a polemical conclusion that the essence of any chemical substance is numerical and probabilistic. Similarly, I shall conclude this chapter by saying that a wave is like a hand of cards and a particle like a bet on the outcome.

Thus the problem of waves and particles ultimately merges into the problem of probability and determinism, the subject of the next chapter.

Chapter Five

Determinism and Indeterminism

The purpose of this chapter is to investigate the psychological conditions under which first determinism and then indeterminism took hold of the modern scientific spirit. I shall also attempt to show how the principles of determinism and indeterminism are bound up with our conceptions of things, space, time, form and function. I believe, in consequence, that these principles are properly studied in all the complexity of their psychological context, surrounded by all the empirical and emotional ambiguity that normally accompanies research on the frontiers of science. When we have done this we shall discover that our psychological attitudes toward the determinate and the indeterminate run almost parallel to our attitudes toward unity and plurality. When, finally, we reach this point, we shall have in hand all the elements we need to state the problem of statistical knowledge.

I

If my aim were to trace the history of determinism, I would be obliged to relate the whole history of astronomy. For it was in the farthest reaches of the heavens that pure Objectivity (a correlative of pure Vision) first took shape. Destiny was governed by the regular motion of the stars. If fate rules human life, it is because each of us is subject to the influence of some star. Hence humankind evolved a philosophy of the starry firmament. This philosophy taught human beings that the laws of physics are absolutely objective and deterministic. Without this important lesson of astronomical mathematics, arithmetic and geometry would probably not have developed in such close association with empirical thought as was actually the case: Terrestrial phenomena are too obviously fluid and diverse to permit, without prior psychological preparation, the elaboration of an objective, deterministic physics. Determinism descended from heaven to earth.

Closer to our own time, it was Newtonian astronomy that lent its rigor to the Kantian categories and its absoluteness to the *a priori* forms of space and time. It was Newton's astronomy that laid the foundations of modern mathematical physics. Astronomical phenomena are in one sense the most objective and fully determined of all physical phenomena.

Astronomy is thus the branch of knowledge most apt to establish fundamental habits and forms of thought that, though not necessarily *a priori* in perception, may justly be called *a priori* in reflection. If we follow the development of astronomy up to the last century, we find that two meanings have been attached to the word *determinism: Determinism* may be taken to be either a property inherent in phenomena themselves or an *a priori* form of objective knowledge. Frequently, it is a subtle shift from one to the other of these two meanings that introduces confusion into philosophical discussions.

Determinism, as I said, has astronomical roots, and it is this fact, I think, that explains why philosophers were so slow to concern themselves with the question of perturbations, errors, and uncertainties in the study of physical phenomena. Not until much later did indeterminism first gain a foothold on this error-ridden fringe of science. Bear in mind that even in astronomy the theory of perturbations is a relatively modern development. Delambre reminds us of Pemberton's view that Newton demonstrated his good judgment by neglecting a few discrepancies of minor importance in his results. It has frequently been pointed out that greater precision of celestial measurement might have hindered the discovery of astronomical laws. In order for the world to appear to be governed by laws, the earliest stated laws had to have a simple mathematical form. Determinism could not have taken hold without a truly elementary mathematics. Simplified empirical data were reinforced by the mathematics, which lent an air of necessity to what was observed. Reasonably accurate observation linked up with reasonably accurate prediction to establish determinism on a firm footing.

For the historian, the problem of the shape of celestial bodies is perhaps even more rewarding to study than the problem of orbital trajectory. For a long time scientists were sure that the celestial bodies must have simple geometrical shapes. When geodesic measurement revealed that the earth is not a perfect sphere but actually rather flattened, there was considerable astonishment. Maupertuis was branded "the intrepid flattener of the Earth."[1] Yet what proof was there that the earth was round, other than that it was possible to circumnavigate it? People were also convinced that the shape of a celestial body has no effect on its motion. This belief was based on a tacit hierarchy that authorized neglect of so-called secondary characteristics. It was this hierarchy of what was important and what was not that lent an air of rigor to determinism.

In short, the mathematical conception of the world was originally inspired by the intuition of simple forms. Because of this intuitive faith in simplicity, people were reluctant to believe that celestial bodies might actually be distorted in shape or follow complex trajectories. Determinism was thus a consequence of the simplicity of the original geometry. The feeling of *determination* was a feeling that a fundamental order exists, a feeling of intellectual repose stemming from the symmetries and certainties inherent in the mathematical analysis.

Once it has been understood that the psychology of determinism derives from attempts to rationalize reality, it be-

[1] Pierre-Louis Moreau de Maupertuis (1698-1759), a physicist and mathematician who was one of the first to measure accurately the length of the meridian and who stated the "principle of least action" in mechanics.

comes easier to penetrate the psychology of *deformation* and *perturbation.* Deformation and perturbation techniques (which did not really come into their own until the nineteenth century) of course depend on the assumption that a fundamentally simple law remains valid even if it yields predictions slightly at variance with the observed data. The discrepancies are interpreted in terms of the basic law. This results in a curious sort of two-stage argument. First, the fundamental solution is given. The perturbation is then introduced as a superficial corrective. Working together, astronomy and geometry protect the determinate character of the phenomenon against all doubt.

If it were possible to forget the philosophical lesson of astronomy and look directly at terrestrial phenomena, it would be obvious that we could never deduce determinism from mere observation. This is a very important point, I think, for it is from direct observation and not reflection or experimentation that we derive our primary psychological categories. Clearly, then, determinism is something that has to be *taught,* and observation has to be corrected by means of experiment. Direct observation could not possibly yield determinism, because *determinism does not constrain all aspects of a phenomenon with equal rigor.* The distinction between *law* and *perturbation* is one that has to be redrawn to suit each particular case. In tracing the evolution of a phenomenon over time, the experimenter singles out crucial moments (*noeuds*) for close scrutiny. Determinism is brought into play at each stage, moving from determinate cause to determinate effect. But one has only to consider the transitional phase between crucial moments to discover processes that are tacitly assumed to be unimportant. To take a very simple example, mixing chalk and vinegar produces an effervescence. The duration of mixing does not influence this result, so it seems reasonable to ignore this factor as extraneous. If, however, one were interested in studying the kinetics of the reaction in detail, one would have to adopt a very different attitude toward the period of time that elapses between the mixing and the onset of effervescence. The evolution of the phenomenon has a history. There is no determinism without deliberate choice, without a deliberate decision to set aside perturbing or "insignificant" phenomena. It quite often happens, moreover, that a phenomenon is insignificant only because one fails to take it into account. At bottom, the essence of the scientific spirit is not so much to observe determinism in action as to determine what the action is, that is, to take precautions to ensure that the preconceived phenomenon occurs without undue distortion.

The simplifying intent that underlies deterministic thinking explains why mechanistic reductionism has enjoyed so much success. *Explanation* and *description* have perhaps never been so far apart as they were in the age of mechanism. If the phenomenological context were restored, it would be apparent that many systems that are assumed to be deterministic are so only to a limited degree, namely, the degree to which mechanics provides an accurate account of their functioning. The epistemological ideal of the age of mechanism was fundamentally reductionist: In order for a phenomenon to be completely determinate, it had to be reducible to mechanical terms.

Taking this argument one step further, I would even maintain that our belief in determinism depends on the possibility of reducing a problem to the terms of *elementary* classical mechanics. On this point Cartan offers the following observation:[2]

> To argue in favor of determinism, in the usual sense of the word, is to argue that the state of the universe at a given moment completely determines its subsequent evolution. Here, of course, we must specify exactly what we mean by a *state* of the universe. The classical mechanics of the material point is deterministic in this sense, supposing we know the position *and velocity* of each material point in the universe at some specified instant of time. . . . What complicates matters somewhat is precisely the fact that the theory of relativity teaches us that time and space are inseparable. To speak of the state of the universe at a given time thus has no absolute meaning. We must actually speak of the state of the universe in a three-dimensional cross section of space-time. This raises still other difficulties, to which Mr. Hadamard has called attention. In reality there is a mathematical determinism and a physical determinism, and the two are not the same. It may happen that the state of the universe in one three-dimensional cross section determines the state of the universe in neighboring cross sections, *and yet the physicist may be unable to detect that this is the case*: This has to do with the fact that a very small change in the state of the universe in the given cross section may entail very large vari-

[2] Elie Cartan (1869–1951), mathematician.

> ations in some neighboring cross section as close as one wishes to the first. The fact that the two states are related is thus completely concealed from the physicist.[3]

In other words, mathematical determinism, which is based on logical deduction, does not coincide as precisely as is sometimes thought with physical determinism, which is presumably based on some sort of causality. To put it another way, causality is not always expressible in unequivocal mathematical terms. It is tantamount to the choice of one state from among a number of possible states. This abundance of possibilities is not based on the choice of a particular instant on the absolute time axis. It is already established in *a unique instant* that can support a number of differently oriented slices in space-time. To speak of a *state* of the universe at a specific *instant* of time introduces two sources of arbitrariness: that of the choice of initial time and that of the choice of initial state at that time.

Simpler examples of arbitrary simplification may also be cited. It has often been noted that mechanics first developed historically as the mechanics of solids. Fluid mechanics came much later. Hence it is hardly surprising that determinism is generally illustrated in terms of relations between solids. In the collision of two solid objects, for example, it is generally taken for granted that the objects that enter the collision are the "same" as those that emerge from it, and that only their "state of motion" changes. Hence it is considered legitimate to reduce the *entire* phenomenon to the analysis

[3] Elie Cartan, "Le parallelisme absolu et la théorie unitaire du champ," *Revue de métaphysique et de morale*, January 1931, p. 32.

of the motion of both objects before and after the collision, as if this took sufficient account of cause and effect. Clearly, determinism in this case depends on a metaphysical assumption that the phenomenology of collision can be divided into two categories, one of objects, the other of motions. Later, we shall examine the validity of this metaphysical dualism. For now I need only remark that intuitive determinism does not hold up very well if we turn from the study of colliding objects to the more complex phenomena of fluid dynamics. Since a flowing liquid is deformed by its motion, it would seem difficult to decide what is the "same" and what is "different"; hence determinism either breaks down or is obscured by various ambiguities. It is only by carrying over into fluid mechanics deterministic ideas from the mechanics of solids that one can regard fluid dynamics as *clearly* deterministic.

Taken together, these general observations suggest that deterministic psychology thrives on a restriction of the allowable range of experimentation. If we compare the way in which astronomy and mechanics are taught with the way in which students develop intuitions on their own, it becomes clear that determinism is a product of selection and abstraction; over the years it has developed its own pedagogical technique. Determinism demonstrates its validity by reference to simplified, monolothic phenomena: Causalism is intimately related to object fetishism (*chosisme*). In mechanics determinism demonstrates its validity by restricting attention to a truncated range of phenomena and to an incorrect analysis of space-time. In physics determinism demonstrates its validity on a hierarchy of phenomena by accentuating the importance of particular variables. In chemistry determinism proves its validity by concentrating on purified substances and listing their properties. Now, notice that the simplified intuitions of mechanics correspond to simple mechanisms; standard physics experiments with billiard balls and pendulums are just simple machines; and pure substances are actually chemical *artifacts*: What is striking in all this is that determinism is demonstrated by artificial technical means. Nature's true order is the order that *we* put into it with the technical means at our disposal. In order to prove that determinism is valid, and especially to teach that it is, one must be careful to simplify: to declare that collisions are elastic, that friction can be neglected, that some laws are important and others are not, that substances are pure. Without such simplifying assumptions the standard demonstrations of physics and chemistry would seem marvelous or fantastic indeed.

To view the problem of determinism as I have just done, as a necessary part of a program of instruction aimed at training the scientific mind, is less misleading than it may appear, for teaching always plays an active role in the formation of the scientific spirit. This would not be the case if science were based on static beliefs or doctrines or on indisputable axioms. For then it would be reasonable to assume that belief in determinism underlies all our thinking and that it, too, is beyond all discussion. It is not difficult, however, to show that determinism is not in fact beyond discussion, indeed that it is the subject of almost daily controversy in the laboratory. Determinism is a metaphysical issue that still weighs heavily on scientific thought, and I therefore think it useful to press my argument still further. To that end, let me introduce a distinction between positive and negative determinism, the meaning of which I shall presently make clear. For the moment I claim only that this distinction is justified by the critical nature of scientific proof. If a critic objects that a particular series of phenomena is not in fact determinate, the only way to answer the objection is to show that given a specified initial state, it is possible to predict such and such a final state (which should be specified as precisely as possible). The more precisely the final state is described, the more convincing the proof will be. There are limits to the degree of precision possible, however. Accordingly, one must acknowledge some measure of ignorance, some slight possibility of variation in the outcome. On the other hand, one can be a good deal more dogmatic in predicting what outcomes *may not* be expected (negative determinism). Here one approaches the absolute, the categorical, the purely determinate phenomenon. One may be absolutely certain that a pocket magnet cannot lift a weight in excess of one kilogram, just as an insurance company can be absolutely certain that none of its clients will live for a thousand years. If any doubts are voiced, one can always resort to exaggerations of this kind in order to restore faith. Thus the foundations of deterministic psychology in a sense rest upon a void. Once faith is reestablished, one can go back to making positive predictions, to saying what the outcome will be (positive determinism): One is then preaching to the converted, who will recognize the truth of the prediction by a sign. But to recognize (*reconnaître*) is not the same as to know (*connaître*). It is easy to recognize what one does not know.

At this point a possible objection crops up. Aren't some signs uniquely revealing or peremptory? In chemistry, for example, the production of a precipitate of a certain color may be a sure sign of a particular reaction. The color of the precipitate is surely a signal characteristic, pointing to the presence of a particular substance. If we pursue the chemist's certainty to its roots, however, we discover that his assured statement that such and such a substance is definitely present is tacitly also an assertion that a series of other possibilities did *not* occur, and that these are precisely the possibilities that might have introduced some ambiguity into the experiment. What is more, when the chemist has identified, say, the presence of some metallic salt, he has said nothing about the purity of that salt and hence has not ruled out the possibility that other metals are present in the original solution as impurities. A slightly more demanding critic, who insisted on a more precise accounting of all the reaction products, might shake the experimenter's confidence in his results. Psychologically speaking, true determinism ultimately rests on negative judgments. Only "nihilistic determinism" can put an end to the interminable polemics inherent in positive proof. Communion of spirits is achieved in negation. The perfect objective union rests upon a species of nonobject.

The purpose of these preliminary reflections on determinism has been simply to analyze the criteria of proof employed. These observations should help us to gauge the validity of determinism by making clear what criteria have to be met in order for a given phenomenon to appear determinate and what initial conditions have to be specified before prediction is possible.

Once these criteria have been made explicit, it becomes clear that causality and determinism are not synonymous and that our psychological belief in causality is not as intimately associated with our psychological belief in determinism as is sometimes thought. Von Mises puts it quite well: "The principle of causality is variable (*wandelbar*), and it subordinates itself to the requirements of physics."[4] I would make an even more general statement, to the effect that the principle of causality subordinates itself to the requirements of objectivity and that in this respect causality is still the fundamental category of objective thought. In fact, the idea of cause and effect was able to establish itself psychologically without the need for complex technicalities of the sort I was forced to rely upon in discussing determinism. Between cause and effect there is a connection that to some extent persists even if the cause and effect themselves are subject to some degree of alteration. Causality is therefore much more general than determinism. Causality is an idea of a qualitative order, whereas determinism is of a quantitative order. When a heated substance expands or changes color, the phenomenon itself teaches us in all certainty what its cause is, but it does not demonstrate that determinism (as defined above) is valid. Indeed, it bears repeating that no positive proof of determinism based on a detailed analysis of microscopic states is possible. The expansion of solids is in fact a statistical phenomenon and as such subject to the laws of probability, just like the expansion of a gas. Drawing such a parallel between solids and gases will ordinarily

[4] Ludwig von Mises, "Uber kausale und statistische Gesetzmässigkeit in der Physik," *Die Naturwissenschaften*, 14 February 1930, p. 146.

provoke resistance in inattentive students. This resistance is a sure sign that intuitions based on the study of solids enjoy an unduly privileged position in the average student's mind.

If the reader has followed me so far in my attempt to distinguish among the fundamental concepts of epistemology, he or she may be willing to accept the following explanation of the reason for the persistent confusion of determinism and causality: namely, that phenomena are subject to what I shall call *topological determinism,* which reflects the existence of functional connections and influences the evolution of general sets (i.e., sets equipped with topologies — trans.) in the same way that the so-called *analysis situs* influences geometrical objects (*analysis situs,* that branch of mathematics which investigates properties that remain unaltered under the continuous group, was the historical precursor of the modern branch of mathematics known as topology — trans.). Topological determinism might well lead to what I shall call *crisis analysis,* a tool for studying the transition from one organic phenomenon to another. If persistent qualities are evident, then quantity may not have much importance. And if certain qualities are characteristic of a particular phenomenon, others may actually be ignored. Causal analysis rests on an obvious hierarchy of qualities, and quantitative determinations are not germane to its purposes.

The forgoing remarks are not merely the idle speculations of a philosopher; they reflect the way in which mathematicians and experimentalists actually think. Scientists do not spend all their time making measurements. They seek first to understand how phenomena are interrelated and often conceptualize such interrelationships in qualitative rather than quantitative terms. It is by associating sign with sign rather than number with number that one begins to unravel a deterministic mechanism. The scientist's faith is rigorous because certain experiments are exempt from the demands of rigor. Actual measurement is relatively rare, and there is therefore room for the sort of topological determinism I outlined above, which could be used to show that a phenomenon remains recognizable despite minor variation in its features.

But it is best perhaps to break off this line of argument here and to have a go at the problem from the opposite direction. In particular, I want now to ask how indeterminism was able to gain a hold over the psychology of the scientific spirit. Let me give away the answer before I start: Scientists were surprised to discover that certain regularities of a relatively precise and demonstrable kind govern even seemingly random phenomena. This will be the subject of the next section.

II

Indeterminism first entered physics in connection with work on the kinetic theory of gases. This theory profoundly and permanently altered the nature of the scientific spirit. Many philosophers have also found the kinetic theory to be a fertile source of ideas. Abel Rey, to name one, has devoted several of his books to its philosophical importance.[5] Hence I

[5] Abel Rey (1873–1940), philosopher of science and director of Bachelard's thesis.

can keep my remarks quite brief.

In my opinion, the deepest metaphysical feature of the kinetic theory is its introduction of transcendental qualities, by which I mean that something that is not a property of any part of a system may be imputed to that system as a whole. Various theories of logic have found this troubling. A quite recent example may be found in an article by Peter Carmichael. Carmichael holds it to be a major error that the behavior of the constituent molecules of a gas is

> unpredictable (i.e., for contemporary physics, indeterminate), whereas the average behavior of a large number of molecules is predictable (i.e., determinate). In other words, the individual object is indeterminate, the class determinate. But this clearly violates the axiom *de omni et nullo* and is therefore self-contradictory. The same conclusion applies to all so-called statistical laws and probabilities in which a property is asserted of a class of objects and denied to the objects considered separately, for otherwise there would be a gap between the class and the objects. . . . The only course open to the scientist is to deny the axiom *de omni et nullo*, or in other words to argue in self-contradictory terms, which is what scientists do when they subscribe to the doctrine of indeterminism.[6]

But in fact it is the metaphysical contradiction that needs to be overcome. Introducing the idea of probability does not eliminate the contradiction entirely but makes it less glaring. The logic of probability is not well understood, but it is clear that the axiom *de omni et nullo,* which is valid for collec-

[6] Peter Carmichael, "Logic and Scientifical Law," *Monist*, April 1932.

tions of objects, cannot be applied in any straightforward way to composite probabilities.

Leaving the logicians' questions aside, then, let us try to get a grip on what indeterminism really is. We start with a collection of objects whose behavior is unpredictable. In the kinetic theory of gases, for example, nothing is assumed about the constituent atoms, which figure only as subjects of the verb *to collide.* Thus we know nothing about the time during which the collision occurs, and since we can't describe the basic phenomenon in any precise way, we can't predict its outcome. We don't know the initial position and velocity of every atom in the system. In other words, we can't specify the initial conditions; hence subsequent states of the system are indeterminable. "Indeterminable" is admittedly not the same as "indeterminate." But when a scientist shows that a phenomenon is indeterminable, he considers himself duty-bound by the canons of scientific method to regard it as indeterminate. The indeterminable teaches him the need for indeterminism.

Now, in order to find a method for determining the course of a phenomenon, one has to assume that it depends on other phenomena, which determine its outcome. Similarly, if one assumes that a phenomenon is indeterminate, then it stands to reason that it is independent of other phenomena. The large number of collisions among the molecules of a gas may thus be regarded as one general phenomenon with many independent constituents.

It is at this stage that the calculus of probabilities comes into play. In its simplest form probability theory is based on the assumption that different events are independent of one another. If there is the slightest dependence, complications arise that usually make it quite difficult to apply the theorems of probability. Thus it was the assumption that molecular collisions in a gas could be regarded as independent that paved the way for the introduction of probability theory into physics.

Now, the psychology of probability remains rather obscure, for it stands in opposition to the whole psychology of action. *Homo faber* has been hard on *homo aleator*; realism has been hard on speculation. There are physicists whose minds are closed to ideas of probability. Poincaré reminds us of Lord Kelvin's strange incomprehension in this respect: "oddly enough, Lord Kelvin was both attracted [to the new theory] and yet, on certain points, refractory. He never grasped the generality of the Maxwell-Boltzmann theorem. He assumed that there must be exceptions, and if one showed him that an exception he thought he had discovered was merely apparent and not real, he set about looking for another."[7] Kelvin, who "understood" natural phenomena in terms of gyroscopic models, found the laws of probability in some way irrational.

Contemporary science has set itself the task of assimilating the laws of probability. Statistical mechanics, still a relatively new branch of physics, incorporates many incompatible basic assumptions. We are still at the stage of "working hypotheses," trying out different statistical methods whose range of application is limited. Bose-Einstein statistics and Fermi statistics may be based on different principles, but each can be useful in a particular area of physics.

[7] Henri Poincaré, *Savants et ecrivains*, p. 237.

Despite the uncertainty of its foundations, statistical physics has already achieved some remarkable successes. These triumphs appear, as I said earlier, to transcend the boundaries between different kinds of physics. Temperature, for example, has been explained in kinetic terms. This apparent transcending of boundaries between mechanics and the theory of heat may be more verbal than real, however. As Eugene Bloch says, "the principle of the equivalence of heat and work is in some sense materialized from the outset by the way in which heat is conceptualized."[8] Still, it is true that one quality can be expressed in terms of another and that even though mechanics is assumed to be the underlying basis of the kinetic theory, the real explanatory power comes from the statistical arguments, not from the mechanics. In the end we must accept statistical proofs. There is room for a positivism of the probable, although it is actually rather difficult to situate it between empirical and logical positivism.

Do not believe, however, that probability and ignorance are synonymous, simply because probability comes into play when we are ignorant of cause. Margenau formulates the issue in a particularly careful way: "There is a considerable difference between the following two assertions: (a) There is an electron somewhere in space, but I do not or cannot know where, and (b) each point is an equally probable location of the electron. What (b) asserts that (a) does not is that if I carry out a very large number of observations, the results will be evenly distributed throughout space."[9] Here we see

[8] Eugène Bloch, *La théorie cinétique des gaz* (Paris: A. Colin, 1921), p. 2.
[9] Margenau, *Monist*, July 1929, p. 29.

that statistical knowledge does in fact have positive content.

It would be equally mistaken to identify the probable with the unreal. Empirical observation can confirm or refute our subjective belief in certain expectation values, which may or may not be expressed in numerical form. Even if our expectations are vague and confused, they are not unreal. It might even be possible to speak of "statistical causality." The following statistical principle, proposed by Bergmann, is well worth pondering: "The event whose mathematical probability is the greatest will occur in nature most often."[10] On this view, *time is responsible for translating probability into reality*, for making the probable real. A static law, based on summing probabilities in one instant of time, is somehow translated into temporal terms. Why this is so has nothing to do with the fact that probability is usually taught by counting the occurrences of some real event, such as the appearance of a head rather than a tail or a black ball rather than a white. For there is the same gulf between *a priori* and *a posteriori* probability as between the *a priori* logic of geometry and the *a posteriori* geometrical description of something real. That the calculated probability should coincide with the measured probability may well be the most refined, subtle, and cogent proof of nature's permeability to reason. The rationalization of empirical statistics probably must proceed by establishing a correspondence between probability and frequency. Campbell, on the other hand, posits a sort of statistical realism inherent in the atom: "the atom is *a priori* more likely to enter a more favored state and less likely

[10] Bergmann, *Der Kampf um das Kausalgesetz in der jüngsten Physik*, p. 49.

to enter a less favored state."[11] In the end, reality, with the help of time, always manages in one way or another to incorporate probability into existence (*finit toujours par incorporer le probable à l'être*).

Whatever one thinks of this metaphysical interpretation, the fact remains that modern science has accustomed us to working with statistical objects, with objects whose attributes are in no sense absolute. And we saw earlier how the teaching of science might benefit from the introduction of examples involving fluids, plastics, and composites in addition to the usual solids. In this way we might discover that apparent low-level indeterminism was in fact treatable at a higher level in terms of what I earlier called general topological determinism, which allows for fluctuations and probability. Phenomena that appear to be basically indeterminate can be statistically combined and treated collectively. And it is on just such collections that the principle of causality can be brought to bear.

Reichenbach has given an excellent account of the exact relationship between the idea of causality and the idea of probability.[12] He shows that even the most rigorous laws require statistical interpretation: "The conditions on which one bases one's calculations are never realized in practice; for example, in calculating the motion of a material point, a projectile, say, we can never know all of the factors that may intervene. Yet we can make excellent predictions thanks

[11] N. R. Campbell, *Théorie quantique des spectres,* cited from the 1924 French translation, p. 100.

[12] Hans Reichenbach, *La philosophie scientifique,* trans. Vouillemin (1932), pp. 26–28.

to the laws of probability, which cover the factors not included in our calculations.''[13] Reichenbach concludes that any application of causal laws to reality involves considerations of probability. He then goes on to propose that the traditional formulation of causality be replaced by the following two propositions:

> 1. Given a phenomenon and certain specifying parameters, we can predict that the probability that at some subsequent time the phenomenon will be in such and state, similarly defined by a specific set of parameters, is E.
> 2. This probability approaches unity as the number of parameters taken into account is increased.

In other words, if it were possible to take account of all the parameters associated with some actual experiment (assuming that the word *all* even makes sense in any actual situation), then we could be sure of subsequent states in full detail, that is, the phenomenon would be entirely predetermined. But this is a limit argument, and deterministic philosophers assume that there are no problems associated with actually taking the limit. They assume that *all* experimental conditions are knowable without even asking if the list of such conditions is countable, that is, without asking if it is even possible to enumerate the ''givens.'' In reality, scientists always assume implicitly that statement one above is correct, and that certain characteristic parameters are adequate to specify the problem fully. These parameters then become the focus of measurement and prediction. Since certain parameters have been neglected, however, prediction

[13] Ibid.

is at best statistically valid. In short, empirical results may always converge toward a determinate outcome, but it is a serious error to define determinism as anything other than a convergent sequence of probabilities. Reichenbach again puts it quite well: "This definition is frequently replaced by a statement of convergence. Such a substitution introduces quite erroneous notions of the nature of causality, including the notion that probability can be excluded from consideration. Such conclusions are just as false as the definition of the derivative as the quotient of two infinitesimals."

Reichenbach goes on to raise an objection of fundamental importance. Nothing proves *a priori* that the probability of any kind of phenomenon necessarily converges to unity. "This suggests," says Reichenbach, "that causal laws can always be reduced to statistical laws." Extending Reichenbach's comparison, we may suggest that there are statistical laws that do not converge to causal laws, just as there are continuous functions that have no derivative. If there are such statistical laws, they contradict Reichenbach's second postulate. Such laws would make it possible to construct a noncausal physics, just as denying Euclid's postulate makes it possible to construct non-Euclidean geometries. Beginning with Heisenberg, work has recently begun on such a nondeterministic physics, which breaks sharply with the dogmatic negativity of classical determinism. Heisenberg's nondeterministic physics actually subsumes deterministic physics by specifying with precision the conditions and limitations under which it is possible to regard a phenomenon for all practical purposes as determinate. Let us therefore turn next to a consideration of Heisenberg's work.

III

The Heisenberg revolution, which has turned physics upside down, has to some extent quieted the conflict between determinism and indeterminism. What Heisenberg did was nothing less than to establish an *objective indeterminacy* in all physical observation. Before his work, it was assumed that errors in measurement of *independent* variables are necessarily independent of one another. Each variable, considered in isolation from the others, could be measured as precisely as one wished. Experimenters believed that they could isolate variables and study whichever ones they chose with ever greater precision. They placed their faith in an abstract ideal, where the only obstacle to measurement was the inadequacy of the instruments available. Then came Heisenberg, who with his uncertainty principle established that errors in the measurement of different variables are objectively correlated with one another. In order to locate the position of an electron, one has to shine upon it a beam of photons. But the collision between a photon and an electron changes the location of the electron as well as the frequency of the photon. In microphysics there is no method of observation that doesn't interact in some way with the observed object. There is an unavoidable interference between method of measurement and object.

Heisenberg was able to express this general observation in the form of a mathematical inequality. If we let q denote the position of a particle and p denote the conjugate momentum, then the error Δq in the measurement of q and the error Δp in the measurement of p are related by the inequality

$$\Delta p . \Delta q \geq h,$$

where h is Planck's constant (in other words, the smaller the uncertainty about location, the greater the uncertainty about momentum, and vice versa). Many other pairs of conjugate variables obey the same fundamental inequality. Position and momentum are most commonly used for teaching purposes, but energy and time satisfy the same relationship. In fact, the Heisenberg inequality is quite general and can be used to relate variables in mathematical constructs that have no intuitive basis in classical physics.

Heisenberg's simple methodological remark has by now been systematized to such a degree that it has become the starting point of all microphysical methodology; indeed, the uncertainty relation is in itself a veritable methodological principle. It forces us to think of microphenomena in essentially dualistic terms. Bohr has pointed out that Heisenberg's inequality occupies the boundary common to both the wave and particle pictures of matter.[14] It is in a sense the hub to which both partial views are attached. Heisenberg summarizes this point as follows: "According to Bohr, this limitation (i.e., the Heisenberg inequality) is readily derived from the principle that all the facts of atomic physics must be intuitively representable both as particles and as waves."[15] Note in passing that the atomic realm is depicted as the place where the two contrary intuitions come together, a fact that will come as no surprise to philosophers familiar with the history of atomism.

The objective dualism implicit in Heisenberg's philosophy

[14] Niels Bohr (1885–1962), Danish physicist and theorist of the atom.
[15] Heisenberg, *Principes*, p. 9, gives an explicit *proof* of Bohr's remark.

naturally has repercussions on a wide range of qualitative relations. For example, Solomon, in his thesis on "electrodynamics and quantum theory" (1931), remarks that if E represents the electric field and H the magnetic field of a (free) electron, then it is just as impossible to determine both simultaneously as it is to determine simultaneously the location and velocity of an electron in an atom. Hence "if we adhere to the Heisenberg principle that theory should make use only of measurable quantities, we must accept the fact that E and H cannot be measured simultaneously." Using these simple remarks and almost no calculation, Solomon is able to derive uncertainty relations among the various components of the electromagnetic tensor and ultimately arrives at the quantum field theory, which has also been discovered, in less direct fashion, by Dirac,[16] Pauli, Jordan,[17] and Heisenberg.

One can hardly fail to be struck by the way in which sound methodology somehow leads to separation of the electrical and magnetic components of the electromagnetic field. Realists were always inclined to impute *reality* to electromagnetism. By linking together the adjectives *electric* and *magnetic* and by coining the word *electromagnetic* to cover two distinct experimental possibilities, realist physicists persuaded themselves that they were laboring under the sign of a real object. Hence they had no reservations about inscribing that field within space itself. They assumed the existence of a physical ether so as to etch the geometric character of fields still more deeply into space. Accordingly,

[16] P. A. M. Dirac (b. 1902), English physicist, major figure in the development of quantum mechanics.
[17] Pascual Jordan (b. 1902), German physicist.

realists are now finding it difficult to give up the classical description of the electromagnetic field in terms of space and time, as quantum field theory is forcing them to do. Yet it is essential to move from intuitive geometrization to "discursive arithmetization" and back to a statistical definition of the field.[18]

Looking at the question from another angle entirely, we can say that relativity actually blurred the distinction between electricity and magnetism inherent in the classical theory of the electromagnetic field. Einstein, in commenting on his new unified field theory, makes the following statement: "The same state of space that appears in one coordinate system to be a purely magnetic field will appear in another coordinate system in motion relative to the first to be an electric field, and vice versa."[19] Thus the ostensible empirical realities that we baptize electric and magnetic fields are actually mere appearances, which can be altered simply by changing our frame of reference.

IV

In light of the foregoing remarks, it seems clear that one of the most important philosophical consequences of the Heisenberg principle is that it sets limits to the assignment of realistic attributes. Any claim to circumvent the uncertainty relations implies that the words *position* and *velocity* are

[18] "Discursive arithmetization" is a literal rendering of the French *arithmétisation discursive*; by this Bachelard seems to mean quantum numbers, though it is difficult to be sure. — Trans.

[19] Cited by M. Metz, "La théorie du champ unitaire de M. Einstein," *Revue philosophique*, November 1929, p. 393.

being used outside the domain in which they were defined (and are definable). The objection that such fundamental notions have universal meaning carries no weight; there is no reason to regard geometrical qualities as primary. All qualities are intimately associated with some relation and hence secondary.

In order to understand why we have such unreasonable confidence in the idea of absolute position, we have only to recall that spatial localization underlies all language; syntax is essentially topological in nature. But it is precisely against such consequences of spoken thought (*la pensée parlée*) that scientific thought must react; in this connection Heisenberg offers the following penetrating remark:

> Bear in mind that human language allows us to form propositions from which no consequences can be drawn, propositions that are in fact completely devoid of substance, even though they produce some kind of image in our minds. For example, the assertion that there may exist another universe alongside our own but in principle having no relation with it whatsoever yields no consequence but does give rise to a mental image of some sort. Of course, such a proposition can neither be confirmed nor denied. We should be particularly circumspect about using the phrase "in reality," because it readily leads to propositions of the type just described.[20]

Another example of the way in which objective designation can be misleading may be seen in the term *atom*: Actually (in the kinetic theory of gases, for example — trans.), we

[20] Heisenberg, *Principes*, p. 11n.

have information not about one atom but about a group of atoms. Obviously, then, our terminology should reflect the fact that we are describing a collective and not an individual reality.

The philosophical conditions of statistical individuation have been analyzed in a very clear way by Chester T. Ruddick. Ruddick constrasts individuation in statistical mechanics with individuation in standard classical mechanics, which assumes that each individual object, that is, each solid, is known by its position in space and time; the laws of mechanics apply to such objects only insofar as they are held to be separate and distinct entities.

> The objects of statistical law, on the other hand, may be subject to an entirely different means of individuation. Their sole distinguishing feature may be their membership in a certain group; they may be "hydrogen atoms" or "men," but not "this hydrogen atom" or "this man." They are distinguished only from objects outside of their own group; not from those within. The law is established on the basis of an assumption that one member of the group is as likely to satisfy certain conditions as any other. All individuating characteristics are wiped away by bringing the individual into the group. Its definition as an individual is its definition as a member of the group. It might be suggested that the same may be said in the case of mechanical law. Newton's universal law, that "all" particles attract each other in a certain way, refers to the members of a group, to points whose defining characteristic is that they have mass. But the derivation and the application of this law depends

> not only upon the recognition of certain points as members of the group, but also upon a consideration of the differences between such points. A particular point behaves as it does in conformity to such a law only because it is particular. Its conformity to statistical law, on the other hand, would depend, not upon its being different from, but upon its being like other such points.[21]

In other words, we must substitute the indefinite article for the definite article and limit ourselves, with respect to the elementary objects of theory, to a finite intension in one-to-one correspondence with the well-defined extension of the object. Henceforth, the way to grasp the real object is through its membership in a class. The search for the properties of the real is a search for the properties of a class.

Many physicists, including Langevin and Planck, have laid stress on the fact that in the new physics the basic objects appear to lose their individuality. About the philosophical importance of this change Marcel Boll has this to say:

> Just as the anthropomorphic concept of force was eliminated by Einstein's theory of relativity, so too must we abandon the notion of object or thing, at least as far as atomic physics is concerned. Individuality is a property of complexity, and an isolated particle is too simple to be endowed with individual qualities. This position of present-day science with respect to the notion of "thing" appears to accord not only with wave mechanics but also with the new statistics and with the unified field theory [of Ein-

[21] Chester T. Ruddick, "On the Contingency of Natural Law," *Monist*, July 1932, p. 361.

> stein], which represents an attempt to synthesize gravitation and electromagnetism.[22]

On this last point, Ruyer writes that it is "a curious parallel (between relativity and quantum theory — trans.) that in Einstein's new unified field theory, which is quite unrelated to quantum theory, the physical individuality of the different points that make up the presumably continuous material or electrical fluid is denied."[23] Ruyer refers the reader to a penetrating article by Cartan, who concludes that "the material point was a mathematical abstraction to which we became accustomed and ultimately ascribed physical reality. It is one more illusion that will have to be abandoned if the unified theory becomes established."[24]

Meyerson has discussed this argument at length.[25] The learned epistemologist does not accept the view expressed by Cartan, because he cannot forget that physicists (when they think as physicists and not as mathematicians) constantly invoke realism in its customary sense. But must we continue to draw a radical distinction between the scientific spirit schooled by mathematics and the scientific spirit schooled by physical experience (i.e., between theoretical and experimental physics — trans.)? If what I have been saying about the remarkable new importance of mathematical physics is correct, then there is nothing to prevent us from saying that the new scientific spirit is schooled

[22] Marcel Boll, *L'idée générale de la mécanique ondulatoire et de ses premières explications* (Paris, 1923), p. 32.
[23] Ruyer, *Revue philosophique*, July 1932, p. 92n.
[24] Cartan, "Le parallelisme absolu," p. 28.
[25] Emile Meyerson, *Réel et déterminisme dans la physique quantique* (Paris: Hermann, 1933).

by mathematical physics (i.e., that mathematical ideas are central to both theoretical and experimental physics — trans.). The problem then becomes one of reconciling rationalism and realism. But a means of doing so presents itself in mathematical physics: Deprived of their individuality, the elements of the real become indistinguishable from one another, but collectively they behave in what may be considered a rational manner, since reason is capable of predicting what will happen. What gives Langevin's position its philosophical force, it seems to me, is that he is concerned with a hypothetical reality (*une réalité postulée*). His method requires denying individuality to this hypothetical reality. We have neither the right nor the means to ascribe individual qualities to elements defined as members of a set. *Elementary* realism is therefore an error.[26] In the realm of microphysics realist thinking must be vigilantly opposed. The situation of science now is reminiscent of when the calculus was first discovered. The physically infinitesimal embarrasses us in much the same way as the mathematically infinitesimal embarrassed the mathematicians of the seventeenth century. We should heed the advice of Eddington,[27] that the modern physicist must exercise "scrupulous care to protect his (fundamental) notions from contamination by concepts borrowed from the other world." Meyerson sees something illusory in this, however: "The concepts of scientific theory must in some way recall those of common sense, or the physicist would have no way of manipulating

[26] Dupréel, "De la nécessité," *Archives de la Société belge de philosophie*, 1928, p. 25

[27] Sir Arthur Stanley Eddington (1882–1944), English physicist and astronomer.

them."[28] Physics has no doubt maintained, in its language at any rate, a more or less realistic flavor, but is it indeed true that this obscure residue of realism actually lends weight to the concepts of physics and influences the goals of research? Isn't it more accurate to say that realistic concepts serve as pretexts for dialectical development, providing temporary working hypotheses that will ultimately be expunged from memory? For example, when a physicist speaks of the *spin* of an electron, is he talking about a *real* rotation? If we were to take a survey, we would find that physicists themselves are divided on this point, with "intuitive" thinkers taking one side and "abstract" thinkers the other. It is striking, moreover, that French physicists use the English word *spin*, as if they wished to impute to their intuitionist English colleagues reponsibility for this choice of imagery. Meyerson, I think, deals only with the problem of imagination, and it is no accident that he bases his view on an argument first suggested by Tyndall, one of the most intuitionistic of all the English physicists.

The epistemological problem of electron spin goes beyond the bounds of intuition for two related reasons:

1. The mathematical idea of *rotation* is the pretext that justifies the use of the spin metaphor. The best proof of this is that rotation is very easily quantized. If physicists were actually imagining the rotation of real particles, with all the extra baggage that such an image would entail, quantization would be much more difficult and complex. Furthermore,

[28] Meyerson, *Réel et déterminisme*, p. 19.

spin is justified when spin-states are composed. With an isolated electron, spin makes no sense. Thus spin is *conceptualized (pensée),* not *imagined.*

2. The rotation of the electron, indeed the electron itself, makes no sense to the ordinary imagination. Bear in mind that *we imagine with our retinas* and not with some mysterious and all-powerful faculty. This point has been brilliantly developed by Jean Perrin.[29] The imagination takes us no farther than sensation. Merely attaching numbers to images is not enough to give us an idea of how small an object is: The path of least resistance for the imagination is not the same as for mathematics. We can no longer think in any way but mathematically. Precisely because our sensory imagination fails us, we must move onto the plane of pure thought, where objects have no reality except in relations. Hence there are limits to what we humans can imagine reality to be; or, to put it another way, there are limits to the image-defined *determination (la détermination imagée)* of the real.

Our conception of microphenomena is therefore not based on the "realist kernel" of the electron concept. This kernel plays no part in our actual "handling" of the concept, which is far more influenced by its "idealist" context. The realist view fails to give sufficient weight to the duality inherent in the idea of substance to which I alluded (citing Renouvier) in my introduction. This duality of *object* is perhaps more

[29] See Jean Perrin, *L'orientation actuelle des sciences* (Paris, 1930), p. 25.

apparent in microphysics than in other areas of science. Let me digress for a moment to say why. In preparing to do an experiment, the physicist does indeed start with the common-sense notion of the real, as Meyerson indicates.[30] In particular, he *designates* what instruments he will use, much as he might designate a table. But when it comes to the intellectual content of the experiment, the physicist does a sudden about-face. He uses his instruments to produce what I shall call instrumental products (electrons, fields, currents, etc.), whose theoretical role is to serve as logical, not substantial, subjects. If traces of substantialist thinking remain, they are to be eliminated as signs of a persistent, and unwanted, naive realism. Meyerson would no doubt object that this persistent realism, "this hydra of a hundred heads that seem invariably to grow back when one thinks they have been severed," is an essential feature of human thought. Why, then, are we repeatedly overcome by frenzy to destroy the hydra? What prescience of our spiritual destiny drives us to *sublimate* our realist ideas? The function of realism ought to be to ensure stability; substantialist explanation is intended to preserve permanence. But in fact, realism has become increasingly fickle. Never before has science had so much disdain as it has now for the objects that it creates. It abandons them without the slightest hesitation.

Accordingly, it seems to me that in the interval between the time when one scientific object disappears and the time when a new reality is constituted, there is room for a nonrealist mode of thought, a mode of thought sustained by its own dynamic thrust. Such an interval, critics will say,

[30] Meyerson, *Réel et déterminisme*, pp. 20–21.

amounts to no more than an ephemeral moment, which hardly counts alongside periods of firm scientific accomplishment, sound explanation, and solid teaching. Yet it is in these brief moments of discovery that we must learn to grasp the decisive turns in scientific thinking. It is by making such moments live again in our teaching that we demonstrate the dynamic and dialectical nature of the scientific spirit. For it is at such times that science must suddenly confront contradictions in the evidence and doubts about its fundamental axioms; it is then that new *a priori* syntheses are proposed, syntheses that, like de Broglie's ingenious theory, create a new reality alongside the old; and it is then that we see those sudden reversals of thought of which Einstein's equivalence principle is one of the clearest examples. Meyerson's whole argument about the persistence of the substantial idea of force crumbles when confonted with a principle of this sort. If we simply recall the fact that, according to the general theory of relativity, an appropriate change of frame of reference is enough to make the gravitational force vanish, the specious character of the notion that this force is "real" becomes obvious.

It does not matter how long realism affords the mind the luxury of intellectual repose; the striking fact is that every fruitful scientific revolution has forced a profound revision of the categories of the real. What is more, realism never precipitates such crises on its own. The revolutionary impulse comes from elsewhere, from the realm of the abstract. Mathematics is the wellspring of contemporary experimental thought.

Chapter Six

Non-Cartesian Epistemology

I

Georges Urbain, a chemist who has been a leader in the careful and systematic application of the new scientific methods, is quick to assert that not even the best of methods can last forever.[1] In his view, every method eventually loses its initial fecundity. There always comes a time when scientists lose interest in searching for the new along old trails,

[1] Georges Urbain (1872–1938), chemist specializing in the study of rare earths.

when science cannot progress except by developing new methods of research. Scientific concepts themselves may lose their universality. In the words of Jean Perrin, "any concept will ultimately lose its usefulness and even its meaning as research departs more and more from the experimental conditions under which it was formulated." Concepts and methods alike depend on empirical results. A new experiment may lead to a fundamental change in scientific thinking. In science, any "discourse on method" can only be provisional; it can never hope to describe the definitive complexion of the scientific spirit.

The fact that sound methods are constantly changing in science is a fundamental feature of scientific psychology. Scientific method is always explicit. Familiarity is no guide when it comes to observation. A method and its application are intimately connected. Even pure theorists must pay constant attention to methodological issues. In the words of Dupréel, "a proven fact is supported not by intrinsic evidence but by proof."[2]

The question therefore arises whether the psychology of science cannot simply be reduced to a matter of *conscious methodology*. It would then be almost a normative psychology, a set of pedagogical principles distinct from scientific knowledge itself. In a more positive light, the essence of scientific psychology would then lie in the reflection whereby experimental *laws* are transformed into *rules* for discovering new facts. This is how laws become coordinated with one another and how deductive thinking is introduced

[2] Dupréel, "De la nécessité," p. 13.

into inductive science. Scientific knowledge accumulates, one might say, without taking up additional room in the mind, and this is one difference between *scientific* knowledge and empirical erudition: Science uses tested methods to filter the facts. This normative element is of course most visible in the psychology of the mathematician, for even though his thoughts are always in some sense "correct," there is a fundamental psychological distinction to be made between what he knows to be true and what he simply conjectures or intuits. But a normative component can also be sensed in the essentially organic conception of things that encrusts logical thought in the world. In any case, experimental testing always begins with what one believes to be logical. Hence the failure of an experiment sooner or later entails a change of logic, a deep change in our thinking. Everything stored up in memory must be reorganized along with the mathematical framework of science itself. There is constant interchange between the psychology of mathematics and the psychology of experimentation. Little by little the dialectics of mathematical thought enters into the empirical realm. Methodological change follows the contours of mathematical argument.

But to get back to Urbain's point, are there methods that are exempt from obsolescence? The answer would seem to be no, provided we judge the matter from the standpoint of scientific research, that is, from the realm in which the assimilation of the irrational by reason never fails to bring about a reciprocal reorganization of the domain of rationality. It is often said, accordingly, that work in the laboratory never follows the prescriptions laid down by Bacon or J. S. Mill. Carrying this one step further, I maintain that there are also reasons to doubt the usefulness of Descartes's methodological dicta. The next section will develop this point.

II

The foundation upon which Descartes erects objective thought is too narrow to accommodate the phenomena of physics. The Cartesian method is *reductive* rather than *inductive.* But reduction distorts analysis and hinders the extension of objective thought, and without extension there is no such thing as objective thought or objectification. What I shall show is that Cartesian method, so useful a tool for *explicating* the world, is inadequate when it comes to *complicating* experience — the true function of *objective research*.

Let me begin by asking what justifies the initial separation of simple natures? To take an example whose generality is especially telling, recall that in microphysics it is objectively impossible to distinguish between object and motion. De Broglie stresses the same point: "Early in the development of modern science, Descartes argued that it was important to explain natural phenomena in terms of figures and motions. The uncertainty relations express the fact that a rigorous description of this kind is impossible, since we can never know both the figure and its motion at the same time."[3] Hence the uncertainty relations should be interpreted as impediments to absolute analysis. In other words, we must de-

[3] Louis de Broglie, *Théorie de la quantification dans la nouvelle mécanique*, p. 31.

fine the basic notions of physics in terms of relations, just as we define mathematical objects by stating the axioms that determine how they relate to one another. Parallel lines do not exist *prior* to Euclid's postulate; they are *ulterior*. Similarly, in microphysics, objects do not exist *prior* to a method of detecting them. Definitions depend on methodology: Tell me how to find you and I will tell you what you are. Simple always means simplified. We cannot use simple concepts correctly until we understand the process of simplification from which they are derived. Unless we are willing to make this difficult epistemological reversal, we cannot hope to understand the real point of mathematization.

We have already seen several times that the phenomena of microphysics are in essence complex. While it was logical for scientists influenced by Descartes to construct the complex out of the simple, modern science tries to read the real complexity of things beneath the simple appearance; it seeks diversity beneath identity and tries to go beyond superficial and summary views. But the opportunity to achieve these goals does not arise of its own accord; it does not lie on the surface or beckon to the observer. Scientists must delve into the very heart of matter, into the fabric of its attributes. It is theory that guides research. What feats of pure thought, what faith in the realism of algebra, was necessary to bring together matter and motion, space and time, matter and radiation! Descartes was able to deny the essential diversity of matter and motion, but modern science, which has found the way to bring matter and energy into collision, has discovered that diversity is everywhere: Simply by changing the parameters of the colliding particles, it is possible to produce different colors, different kinds of radiation, different amounts of heat. Matter is no longer just an obstacle that repels moving particles. It transforms them and is itself transformed. The smaller the "grain" of matter, the greater its substantial reality; as we observe smaller and smaller volumes, the structure of matter becomes deeper and deeper.

Hence in theoretical physics even more than in experimental physics there is a need for synthetic *a priori* judgements. An organic conception of microphysical phenomena, bringing together at a deep level all of the fundamental notions of the subject, has therefore become increasingly essential. The main goal of contemporary physics, as we have seen, has been to effect a synthesis of matter and radiation. Underlying this physical synthesis is a metaphysical synthesis of object and motion. The synthetic judgment involved is quite difficult to formulate, because it runs strongly counter to the analytical habits formed in the course of ordinary experience, for which there is no difficulty at all in dividing phenomenology into two separate categories, one static (the phenomenology of the object), the other dynamic (the phenomenology of motion). But it has now become essential not to isolate the various aspects of a phenomenon; in particular, it is important to do away with the notion of an object *at rest*. In microphysics, matter at rest is an absurdity, because matter exists for us only as energy and sends us messages only by way of radiation. What, then, is an *object* that can never be examined in its static condition? Nothing suitable for physical analysis. The physicist seeks instead to combine position and motion in his equations, in which the variables describing the one and the other are said to be *conjugate* (the term was taken originally from the Hamiltonian formalism for classical mechanics — trans.). Of course his use of these conjugate variables is still guided by ordinary intuition, and this might lead one to believe that all that is going on is that two simple notions are being combined. But a closer look at mathematical physics gives the lie to this erroneous assumption. For it becomes clear that the conjugate variables enter into the calculations in an indirect manner, and that what is called momentum need not correspond to our original, intuitive idea of momentum. The key parameters are actually derived from a general mathematical formalism. In other words, the usual concrete description of the problem is replaced by an abstract mathematical description. This mathematical description is not clear from its elementary constituents but only from the final equation, whose synthetic value is manifest.

In raising the idea of a non-Cartesian epistemology, my intention is not to condemn Cartesian physics or mechanism (whose spirit remained Cartesian) but rather to criticize the doctrine of simple and absolute natures. The new scientific spirit has profoundly altered our understanding of intuition. Intuition is no longer direct and prior to understanding; rather, it is preceded by extended study, the result of which is to develop a fundamental duality in our thinking. All our elementary ideas are doubled, ranged alongside complementary ideas. Intuition must henceforth involve a choice. Hence scientific description cannot help being fundamentally ambiguous, and this fact raises problems for the immediacy of intuition on which Cartesian judgment relies. Descartes believed not merely that absolutes exist in the objective world but also that they are wholly and directly known to us. Our clearest ideas, in fact, pertain to these absolutes, in Descartes's view. This is so because simple things are indivisible. We see them whole because we see them as separate objects. Clear and distinct ideas are wholly free of doubt, and by the same token simple objects are wholly free of relations with other objects. Nothing could be more anti-Cartesian than the slow change that has been brought about in our thinking by the progress of empirical science, which has revealed a wealth of information never suspected in our first intuition. Einstein's discovery of relativity, for example, has revealed a world of such richness and complexity that the inadequacy of Newtonian physics soon became apparent. Similarly, de Broglie's wave mechanics *completes* (in the full sense of the word) both classical and relativistic mechanics.

But let us assume for a moment, with Descartes, that the elements of the real are truly and integrally given. Can we at least say that the Cartesian construction that unites them is truly synthetic in form? The answer, it seems to me, is that the Cartesian inspiration remains analytical even here, for Descartes believed that no construction is clear to the mind unless the mind knows how to take it apart. He tells us that when we confront something composed of many elements, we must always look for the simple parts, the basic components of the system. Never is a composite idea comprehended in terms of its synthetic value. Descartes never pays heed to the reality of the complex, to the emergence of qualities in the whole not evident in the parts. Rather than accept the complex idea of energy, for example, he prefers to go against the intuition of his own senses and favors a thoroughly reductionist account developed by "intellectual intuition." Nor would he accept that curved trajectories are truly elementary. The only true motion, for Descartes, was simple, uniform, rectilinear motion, the only motion of which the mind can form a clear idea. In thinking about motion along an inclined plane, he was reluctant to assume that the speed of the moving object varies continuously, because speeds in his view should be separate entities, simple and distinct *elements* of a well-defined fall.

Once again, let me compare these tenets of Cartesian epistemology with the contemporary scientific ideal of complexity. Modern science begins with synthesis. Its basis is a complex amalgam of geometry, mechanics, and electricity. It develops its arguments in the context of space-time. It puts forward innumerable sets of axioms. Epistemologically, it relies for clarity on combining ideas rather than on trying to understand individual objects in isolation. In other words, instead of intrinsic clarity it relies on what I shall call operational clarity. Relations do not exemplify objects; objects exemplify relations.

Of course the fact that the epistemology of contemporary science is non-Cartesian should not blind us to the importance of Cartesian thought, any more than non-Euclidean geometry should blind us to the limpid organization of Euclidean thought. But these different ways of organizing our ideas should suggest a general truth about systems of thought that lay claim to totality. *Completeness* should henceforth be regarded as a *de jure* rather than a *de facto* issue. Whether a system is or is not complete can be judged in ways that have nothing to do with the simple enumeration of possibilities. Contemporary science does not enumerate, it theorizes. It does not count up its wealth but proposes new ways of amassing further riches. We must pay close attention to whether or not our knowledge is complete and remain on the lookout for occasions to extend it, to follow one or another avenue of dialectical development. We want to be sure that we have listed all the variables relevant to any particular phenomenon. When we ask ourselves if we have in fact discovered all the degrees of freedom of a system, we are obviously asking a theoretical and not an empirical question: Has anything been left out of the theory? We may find that something was overlooked in our original intuition. It is *theoretical oversights* that we fear, for it is obvious that physicists and mathematicians don't *forget* anything — they simply leave things out.

What we discover when we study physics is that experiments, however important as methods of proof, are always simplifications, selections, examples, or extensions of theory. Intuitive ideas are made clear in a discursive manner, by progressive illumination, by illustration in a series of examples that bring one or another notion into clearer focus. Dupréel elaborates on this point as follows:

> If, by an intellectual act, I posit a simple truth, a second act is needed in order to see what it is good for (*s'en rendre compte*). Generalizing this observation, we see the error of those who believe that necessary and unconditional truths, duly regarded as such, can both at the same time be usefully applied. Once an axiom is posited, a second act is always necessary to establish its application, that is, to establish the circumstances under which it may be invoked. How was it that Descartes and all the other champions of self-evident necessity failed to perceive that the crucial moment comes not when one hangs a hook on a wall,

> no matter how solid that hook may be, but when one finally attaches the first link of a chain of deductions to that hook? However irrefutable your *cogito* may be, I am waiting for you to draw from it some [concrete] conclusion.[4]

It is impossible to give a better statement of the discursive nature of clarity and of the identity that exists between evidence and varied application. In order to measure the epistemological value of a fundamental idea, we must always look to induction and synthesis. For it is there that we discover the importance of the dialectical process that discloses variety within identity and that clarifies by completing our initial intuition.

III

The reader may be willing to grant that Descartes's rules of thought do not answer the multifarious requirements of research in either theoretical or experimental science and yet still object that those rules retain some pedagogical value. Here again, however, I must insist on the gap between the true spirit of modern science and the simple impulse to order and classify. If I may draw an analogy between science and religion, what is needed is a distinction between ''regular'' and ''secular'' science (in Catholicism, the regular clergy are bound by the *regula* or rule of an order and live together in communities, whereas the secular clergy work in the world, or saeculum, ministering to their flock — trans.): Regular science is the science practiced in the research laboratory,

[4] Dupréel, ''De la nécessité,'' p. 14.

while secular science finds its disciples among the philosophers. If the issue is one of teaching students the importance of taking orderly notes, writing clear reports, making precise conceptual distinctions and thorough observations, then nothing is more valuable than the teachings of Descartes. They amply instruct the student in the use of those careful, objective methods that give historical or literary accounts their rightful authority, even as the mathematical and physical sciences are expressing themselves with greater reserve. And in any case it is scarcely conceivable that a physicist could ever violate Descartes's rules. None of the major revolutions in physics has resulted from correcting an error of this kind.

Furthermore, it is apparent that in the context of modern culture there is nothing very startling about Descartes's rules. Not one reader in a hundred of Descartes's *Discourse on Method* has an intensely personal intellectual reaction to that once path-breaking work. Forget the historical charm of the *Discourse*, forget its winning tone of innocent and naive abstraction, and the book has little to offer but common-sense advice — a rather dogmatic, if uncontroversial, set of rules to guide the intellectual life. To the physicist it will all seem obvious. Descartes's rules say nothing about what precautions to take in order to ensure that a measurement will be precise; they offer nothing to calm the anxieties of contemporary scientists. On the other hand, they do prevent teachers from introducing the simple paradoxes that can be so useful in teaching even elementary science. Based on my own experience of teaching basic physics and philosophy, I can say that it isn't easy to interest young minds in the Cartesian method. The problem is that we are now at a stage in human intellectual development where we face a real, and useful, crisis in our understanding of the physical world, but there is no corresponding crisis in our philosophical culture.

Cartesian skepticism, which should be the point of departure for any attempt to teach metaphysics, is not easy to get across. As Frost says, it is a far too solemn affair (*ein sehr feierliche Gebärde*).[5] It is quite difficult to concentrate a youngster's attention long enough to make him perceive the value of doubt. Suspension of judgment prior to objective scientific proof (which is characteristic of the scientific spirit) and clear understanding of the axiomatic meaning of the principles of mathematics (which is characteristic of the mathematical spirit) reflect a less pervasive skepticism than Descartes's, but a skepticism that for that very reason is clearer and more robust. Psychologically speaking, this more limited skepticism, a prerequisite of all scientific research, is still useful. It is an essential, not a transitory, feature of the structure of science.

IV

The time has now come to leave these generalities of method behind and to attempt, by examining several specific scientific problems, to show the new epistemological relationship between simple and composite ideas.

There are no simple phenomena; every phenomenon is a fabric of relations. There is no such thing as a simple *nature*, a simple substance; a substance is a web of attri-

[5] Walter Frost, *Bacon und die Naturphilosophie* (Munich, 1927), p. 65.

butes. And there is no such thing as a simple idea, for as Dupréel has pointed out, no idea can be understood until it has been incorporated into a complex system of thoughts and experiences. Application is complication. Simple ideas are working hypotheses or concepts, which must undergo revision before they can assume their proper epistemological role. Simple ideas are not the ultimate basis of knowledge; after a complete theory is available, it will be apparent that simple ideas are in fact simplifications of more complex truths. If one wants to understand the dialectics of the simple and the complete, nothing is more instructive than to consider experimental and theoretical work on atomic structure and spectra, an almost inexhaustible source of epistemological paradoxes. To take an example that we shall soon be studying in some detail, it is fair to say that in certain respects an atom with several electrons is simpler than an atom with only one electron: the more complex the organization, the more organic its complexion. The curious quantum-theoretical notion of degenerate states sheds a new light on certain supposedly simple phenomena. The epistemological implications of such problems are worth pursuing.

Let us consider, therefore, how physicists approached the problem of atomic spectra. The first spectrum that scientists were able to decipher was of course that of the hydrogen atom. The lines of the hydrogen spectrum are more clearly organized into distinct series than the lines of any other spectrum, and it was a part of the hydrogen spectrum that was first described in mathematical terms, in the formula for the so-called Balmer series. What is more, the hydrogen atom itself was seen as having a particularly simple structure, consisting of a single electron revolving around a single proton. Thus the problem of hydrogen was thought to be simple in two respects: (1) A simple mathematical formula describes its spectrum, and (2) it was possible to give a simple intuitive picture of its structure (the so-called Bohr model — trans.).

Physicists then tried to understand more complex atoms in terms of information derived from the study of hydrogen, which constituted a kind of "working phenomenology." Here, science adhered to the classical Cartesian ideal. The mathematical and intuitive approaches to the study of complex atoms are therefore interesting to consider.

Mathematically, it became clear that, allowing for a suitable correction coefficient, an analogue of the Balmer series can be found in the spectra of other atoms besides hydrogen. The coefficient turned out to be the square of the atomic number (i.e., the number of protons in the nucleus of the atom — trans.). This coefficient did not appear explicitly in Balmer's original formula because the atomic number of hydrogen is one. For a time, then, the Balmer formula was held to be valid for all atoms: It was, physicists believed, the simple and general law of atomic spectra.

Improvement in the techniques of spectroscopic measurement eventually led to corrections in certain parameters of the formula, however. These spoiled the elegant simplicity of the original mathematics. Yet these corrections, essentially empirical in origin, appeared to leave the functional role of the various terms in the formula intact, so that the rational basis of the theory still seemed valid. The empirical results could be explained, it was believed, as *perturbations* of a general law. Two-stage processes of this sort are actually rather common in the history of science In the first, more feverish stage, the general law is established. This is followed by a second, more relaxed stage, in which various complexities, at first ignored, are interpreted as perturbations of the simple, general law. This second stage generally lasts for some time. The whole process is a fundamental feature of a characteristic psychological structure, in which there is a sharp division between the clarity of theory and the inevitable murkiness of reality, between what is lawful behavior and what is irregular, and indeed, as is all too readily assumed, between what is rational and what is irrational. These oppositions mark the boundary between intellectual courage and intellectual torpor. Isn't the theorist's work done, some will say, when the broad outlines of a phenomenon have been worked out? Do details, subtleties, fluctuations, really matter? Isn't it enough to "interpret" them in terms of the law or to declare them to be marginal disturbances? A strange dialectic, and a strange peace of mind.

The temptation to achieve a clear result in a short period of time is sometimes so great, however, that the scientist will persist in applying a theoretical model inappropriate to the problem at hand. If we imagine that we see a unicorn in the clouds, the wind may blow for a long while, stretching the fantastic animal this way and that without eliminating it from view altogether; but if someone should interrupt our reverie, the unicorn is likely to vanish in an instant. If enough perturbations accumulate, it may become necessary to take a fresh look at a complex problem. This was precisely what took place in the mathematical theory of spectra: The introduction of matrices made it easier to deal with a large number of terms. I shall have a bit more to say about the complicated mathematics involved in a moment. But first I want to show that there was a parallel increase in complexity of the associated atomic "models."

What happened to the mathematical formulas also happened to the "images" that illustrated those formulas. Again, an effort was made to replace the electron trajectories in the original models with more complex, perturbed trajectories. Discrepancies quickly emerged, however, because the helium atom — simple as it is with its two electrons and its nucleus — proved to be quite difficult to analyze. Research therefore focused on what was called the hydrogenoid character of certain elements in either the normal or the ionized state. And investigation did indeed turn up Balmer-type series in the spectra of ionized helium, alkaline metals, and ionized alkaline-earth metals (all of which possess a single electron in the outer shell, this being the property referred to as hydrogenoid — trans.), from which it was inferred that the spectral picture is basically the same as for hydrogen, so long as there is a single electron revolving around a more or less complex nucleus. In other words, the spectrum of an atom depends almost exclusively on the outer electron. It appeared, then, that the original simple picture was finally vindicated, and that the general law was now understood.

But complexity would ultimately exact its revenge: Not only was it a mistake to look for an artificial hydrogenoid character in other elements, but it also turned out that this so-called hydrogenoid character was not really a simple character at all. Indeed, if I may anticipate the ultimate results, hydrogen turns out to be no simpler than the hydrogenoids, and its pseudosimplicity is actually a misleading sign that obscures what is really going on. Paradoxical as it may seem, it is nonetheless true that if we really want to understand the hydrogen spectrum, we should first study the more complex hydrogenoid elements. In short, the only way to form a correct idea of the simple is first to study the complex in depth.

The problem is that for quantum-theoretical purposes the hydrogen atom doesn't know how to count: Bohr's model seems to allow room for only a single quantum number. Léon Bloch puts it quite well: "The spectrum of hydrogen is in fact a *degenerate* alkaline spectrum, that is, a spectrum that fails to reflect differences in the value of *l*,[6] where *l* is of course the "azimuthal" quantum number (corresponding to classical angular momentum — trans.), which in the nondegenerate case gives rise to the double periodicity observed in the spectra of alkaline metals. And that is not the end of the story. It is possible to assign still a third quantum number to the so-called optical electron of the alkaline metals, and this leads to the prediction of still a third periodicity in the corresponding spectra. "It is interesting," writes Bloch,

> to investigate whether traces of this triple periodicity exist in the hydrogen atom itself when we regard it as a degenerate instance of the alkaline metal case. There are, predictably, substantial difficulties in carrying out experiments to do this. For in lithium, the first true alkaline metal, the doublet structure is so subtle that it took very special conditions to render it visible. For hydrogen the expected doublet struc-

[6] Léon Bloch, "Structures des spectres et structure des atomes," *Conférences d'Actualités scientifiques et industrielles*, 1929, pp. 200, 202.

> ture is finer still. Nevertheless, the resolving power of present-day interference spectroscopes is so great that it has been possible to detect, beyond any doubt, the fine structure of the lines of the Balmer series, and in particular the red line Hα. . . . The decomposition of the lines of H I and He II into extremely tight multiplets, whose structure is the same as that of the alkaline multiplets, shows that there is no essential difference between the hydrogen and hydrogenoid spectra. . . . Thus we see that the simplest of atoms is already a complicated system.

A possible objection to this line of argument is the following: If Peter resembles Paul, then Paul resembles Peter, and if hydrogen is spectroscopically similar to the alkaline metals, then the reverse is also true. But this objection misses the point that a shift has taken place in the *basic image* of the phenomenon, a shift that entails a total transformation of the underlying phenomenology. Careful examination of the sequence of events leads to the following conclusions: The hydrogenoid image is not imposed upon the alkaline metals, but rather the reverse. At the Cartesian stage of research, when investigators were still trying to work from the simple to the complex, the results were stated in the following form: The alkaline metals have a hydrogenoid spectrum. But at the non-Cartesian stage, by which time research had achieved a complete picture of the phenomenon and turned back to the case of hydrogen as a simplified or degenerate instance, the results could be summed up as follows: The hydrogen spectrum is an alkalinoid spectrum. In order to give a detailed account of the spectroscopic data, the more complicated spectrum (here that of the alkaline metals) had to be treated first. It was this spectrum that first opened experimenters' eyes to the fine structure of the spectral lines. No one would ever have looked for the doubling of lines in the hydrogen spectrum had such doubling not already been found in the alkaline spectra.

A similar problem arises, as we shall see shortly, in connection with the hyper-fine structure of the hydrogen spectrum. It is clear that study of the hydrogen spectrum alone would never suggest the need for still more careful investigation. The Balmer formula has no clues to offer experimentalists. Nor does the Bohr model suggest possible theoretical reasons for additional spectral periodicities. The only reason to consider the angular momenta of the nucleus and the orbital electron in hydrogen is that this procedure proved successful in dealing with more complex, that is, more organic, atoms.

It is not simply in regard to theory and intuition that the hydrogen atom turns out to be deficient: It also leaves much to be desired from the experimental point of view. Powerful instruments and extreme precision are required to detect, in the relatively unsophisticated hydrogen atom, certain phenomena that are more easily detected elsewhere. The most apparent features are not always the most characteristic, moreover; we must be careful not to fall into the trap of a "positivism of the first glance." If we fail to heed this caution, we risk mistaking a degenerate case for an essential truth.

Thus, even though it is historically true that the study of the hydrogen atom was instrumental in the development of spectroscopy, hydrogen is far from being the best starting place for further inductive theorizing. The theory of alkaline spectra is an inductive extension of the theory of the hydrogen spectrum. But then the consequences of the new theory for the hydrogen atom have to be deduced from the results for the alkaline metals. Then further induction is necessary — induction is always necessary. And in this case it leads to the discovery of additional structure in the hydrogen spectrum; or, better still, it leads to the *production* of additional structure by powerful, artificial means.

Thus far, we have followed the interplay of the simple and the complex in only one case, that of the hydrogen-hydrogenoid spectrum. But if the hydrogen model proved insufficient, it seems reasonable to expect that the hydrogenoid model too will ultimately be found to be an artificial simplification rather than a definitive theoretical model. And indeed, the model does prove less and less adequate as one moves from the first to the eighth period of Mendeleev's periodic table. The spectra of bismuth and lead bear little resemblance to hydrogenoid spectra, and the spectrum of iron is completely indecipherable in terms of the hydrogenoid pattern.

How does science react to such a failure? By declaring that reality is hopelessly complex and fundamentally irrational? Such an answer, presuming as it does acceptance of defeat, hardly does justice to the courage of modern science. The scientist's response is rather to pursue his education in the theoretical and experimental study of complex phenomena. Theoretically, there is reason to hope that wave mechanics will provide the means to calculate, *a priori*, the terms of spectral series unrelated to the Balmer series (though numerous approximations may be necessary). On the experimental side, clarity should come from study of the hyper-fine structure. Just as the study of the fine structure of the alkaline spectra helped to clear up the degenerate structure of hydrogen, so, too, may the hyper-fine structure of complex spectra such as that of bismuth help to introduce new models into general spectroscopy. In Bloch's words, "the more spectral analysis was refined, the more allegedly simple structures seemed to decompose. The hyper-fine structure, like the fine structure, may prove to be not the exception but the rule."[7] It would be impossible to overestimate the importance of this observation. It is indicative, I think, of nothing less than a Copernican revolution in empiricism. The very idea of perturbation seems destined for eventual elimination. No longer will there be simple laws and perturbations but complex, organic laws, occasionally affected by various "viscosities" or "vanishing terms." The old ideal of the simple law was reduced to nothing more than the simplicity of an example, a truncated truth, a schematic outline, a blackboard sketch. These simplified images still have their usefulness for teaching purposes, because historically teaching has relied on suggestive, enticing examples. But the ease of teaching by means of such devices, our easy confidence in the known, our tranquil acceptance of established systems, has been dearly bought: The easy way is always dearly bought. The danger is that the student will mistake the scaffolding for the finished edifice. Profound knowledge is finished knowledge, and the finished structure of the old physics lay in the details of perturbation theory, in the sophisticated approximations to exact solutions. That was

[7] Ibid., p. 207.

where the equation between noumenon and phenomenon became real, and where the noumenon revealed the technical factors that impelled its further development. But now the static duality of the rational and the irrational has been supplanted by the dialectics of active rationalization. Theory complements experience. Exceptions are eliminated from above, as it were, by adding attributes and functions to account for the accumulation of accidental facts.

That complete theory should take precedence over sophisticated observation is strikingly clear when we look at an experimental observation such as the Zeeman effect, which was responsible for a long series of theoretical developments. The Zeeman effect involves the splitting of spectral lines under the influence of a magnetic field. This observation raises the following question: ''Might not such a separation exist in latent form in the absence of a magnetic field?''[8] This is just another way of saying that the real structure of the atom should be interpreted in terms of various ''principles of possibility''; it reflects the scientist's confidence that *compossible* phenomena [phenomena that vary together] are the first, and eminently rational, sign of something real. Thus we begin to think in terms of *prior* structure (prior, that is, to experiment — whatever it is in the atom that interacts with the magnetic field to produce the Zeeman effect is a ''real'' part of its structure — trans.); we begin to think in terms of *projects* for constructing the structure of the atom, *plans* for getting at its reality. In other words, we begin to fashion theoretical molds to shape our experimental technique.

[8] Ibid.

In a similar vein, we may ask if it really is as absurd as it may seem to wonder what role the Pauli exclusion principle might play in the hydrogen atom. What exactly does this question mean? The Pauli principle is absolutely general. It states that no two electrons in the same atom can have identical quantum numbers. How is this principle to be interpreted in the case of hydrogen, which has only one electron? We can of course opt for simplicity, rejecting the lesson of the Pauli principle (which is based on the study of more complex atoms) on the grounds that there is only one opportunity for quantification. This choice leads to a simplified picture of hydrogen and limits the experimental possibilities. Should we instead imagine phantom electrons that would provide opportunities for more complex quantum-mechanical analysis? Again, the problem is the same: how to count with a defective abacus, how to interpret the law of large numbers when we are dealing with numbers that are small, how to recognize the rule and all its exceptions when dealing with but a single, and manifestly exceptional case? Stated in more general terms, the question is this: How can the simple illustrate the full? Here we have the hydrogen atom, stoichiometrically the simplest element of all, just as the amphioxus is the simplest vertebrate. There is no doubt that hydrogen combines the two electrically simplest forms of matter, the positive and the negative charge. How are we to untangle the skein? Why not complete the tangle by pushing matter's powers of composition to the limit? Won't various functions be clarified once we have seen their full scope? What holds matter together may become all the more obvious once we have created a tightly woven fabric of particles, increasing to the maximum possible extent the number of relations, functions, and interactions involved. The free electron teaches us less than the bound electron, the atom less than the molecule. Yet we must take care not to push such composition too far. We must remain in the realm where composition is still organic if we wish to understand the equation of the complex and the complete.

We have in fact just recently entered the "age of the molecule," after long years devoted to the theory of the atom. To persuade ourselves of the importance of this new age we have only to think back one hundred years: The artificial nature of the old concept of molecule then becomes quite evident. The definition of *molecule* common at the time depended on an obviously artificial distinction between physical and chemical phenomena. A molecule of a particular substance was the smallest unit of that substance that could be separated out by physical means, whereas an atom could be separated from a molecule by chemical means. Viewed as a composite, a molecule was little more than an amalgam of atoms. All the chemical functions were thought to be inherent in the elements, that is, in the constituent atoms. In keeping with the teachings of realist metaphysics, scientists believed that the categorical attribution of properties to elementary substances had explanatory value. Little by little, however, people began to realize that it was not at all obvious why all properties must inhere in the simple elements, and that some properties might be attributes of the molecules themselves. Take one example: the concept of valence. Chemical valence, a scientific concept that was originally a rationalization of the latent substantialist notion of affinity, is now judged to be something that has no precise meaning apart from actual compounds. As Cabrera says, "valence is something more complex than once thought, which has to do with the stability of the new dynamical configurations of the outer electrons resulting from the mutual perturbations of atoms in contact with one another. It is clear that the details of this configuration and the degree to which it is stable depend on the structure of the atoms involved, so that, strictly speaking, valence is not a property of each isolated element but of all the atoms in a compound."[9] In other words, affinity depends upon communion. To enter into a compound is to "come to terms" with all the other elements in that compound. It is misleading to think that an element is so "different" that it cannot possibly enter into an association with other elements — just as misleading as to think of "difference" as an impediment to human associations. Hence there is no reason to pursue the study of the simple element as such, the being-in-itself, since properties are a product of the compound, of the relation.

The propositions that I am setting forth may seem dangerous in that they run counter to the usual practice of dogmatically asserting the basic notions of a discipline. But the very idea that such basic notions exist is in some respects contradictory. Empirical notions derived from ordinary experience have to be revised and modified repeatedly before they can be of any use to microphysics, which defines reality by *inference* rather than *discovery*. Non-Cartesian epistemolgy is thus by essence and not by accident in a constant state of crisis. Let me try once again, therefore, to demonstrate the way in which modern science attempts,

[9] Cabrera, "Paramagnétisme et structure des atomes combinés," *Activation et structure des molecules*, 1928, p. 246.

by focusing on complex cases, to fashion its preliminary ideas into solid, organic definitions.

In the nineteenth century scientists believed, just as firmly as Descartes two centuries earlier, that the rational foundations of mechanics were unshakeable. Even obscure notions like force were thought to designate definite objects, without conceptual mediation. Other notions, such as work and energy, were then derived from these elementary notions: Work (or energy), for example, was defined as the product of a force and the distance thorugh which it operates. This manner of construing energy conformed to the then prevalent analytical and Cartesian ideal. Note in passing that the absolute separation of space and time lent itself to such an intuitive, analytical approach, even though a good many philosophical difficulties persisted, for example, the discrepancy between the static and dynamic conceptions of force. In exploring such difficulties, scientists discovered any number of obscure points in the original definitions of such concepts as force, work, energy, and power and realized that various ambiguities had been carried over from the prescientific era. Indeed, it was realized that there can be no precise definition that fails to relate force to one of its fundamental properties, which is the ability to do work. From a modern standpoint, the essential relationship between the two notions, force and work, is obvious. The reciprocal relationship between force and energy has become increasingly apparent. What will the basic notion ultimately be? It is of course too early to give an answer to this question. Quantum theory may yet end the debate in an unexpected way by disclosing new principles on which to base the mathematical definitions of hitherto empirical notions. In fact, if we look carefully at London and Heitler's ideas concerning the possible interactions of two hydrogen atoms, it becomes clear that in "microenergetics" the tendency is to view force as a derivative notion, a secondary phenomenon a kind of conventional special case. Without defining energy in terms of more or less hypothetical forces, Heitler and London begin by writing energy equations for the two-atom system. By then applying the Pauli exclusion principle to the system, they discover that there are two possible energy states. If when the distance between the two nuclei is decreased the energy of the system increases, then we *say* that the nuclei repel one another; if, on the other hand, the energy of the system decreases, then we *say* that the nuclei attract. Thus what had seemed to be clearly phenomenal characteristics of the problem, such as whether the nuclei attract or repel one another, here become matters of definition. The idea of force has no absolute basis, and in this analysis it ceases to be a primitive notion. Let us carry this line of argument one step further. It is clear that attraction can occur only between two atoms differentiated according to the Pauli principle, and that the possibility of elastic collision, which used to be explained as the result of a repulsive force inherent in the two nuclei, is an attribute of the system in which the two atoms are not differentiated according to the Pauli principle. Thus it seems that attraction occurs between "different sets of quantum numbers" and repulsion between "like sets of quantum numbers." The force that results from this mathematical analysis is a mere phantom of the force that realist metaphysics imagined as the basis of the notion of energy. On this view, mechanical force becomes as much of a metaphor as the force of sympathy or antipathy. It is a property of a composite, not of the elements that enter into that composite. Mathematical intuition, with its concern for completeness, replaces empirical intuition with its arbitrary simplifications.

I believe, in short, that science is beginning to base its reasoning on tentative, complex models, and that the idea of simplicity is reserved for specific, and always provisional, purposes. The desire to leave the canons of explanation open reveals the receptive psychology of modern science. Some new combination of experiments may lead to alteration of the fundamental postulates. As Cabrera wrote in 1928, "we are not yet . . . in a position to say whether quantum mechanics, created to interpret the radiation of isolated atoms, is an adequate tool to explore the far more complicated problem of molecular dynamics. It is possible and, as I believe, highly probable that a new assumption will have to be added to those already in place. In any case, we must keep our minds open to this possibility."[10] Thus mathematical physics is subject to the same anxiety as geometry: There is always the fear that a new postulate may split the science in two. The attitude that we must remain always in doubt of knowledge that had seemed certain in the past extends and indeed transcends Cartesian precaution and is truly worthy of being called non-Cartesian, as long as we remember that non-Cartesian philosophy complements Cartesian philosophy without contradicting it.

Similarly, as I have tried to show in my book *Le pluralisme cohérent de la Chimie moderne (The Coherent Pluralism of Modern Chemistry)*, chemistry has finally succeeded in es-

[10] Ibid., p. 247.

tablishing itself on a firm rational and mathematical footing by systematically extending its pluralist outlook. The way to rationalize the world of matter, it turns out, is to complete it.

Thus it is awareness of totality that animates work in both mathematical physics and pure mathematics. This accounts for the importance of group theory in both disciplines. The mind cannot rest easy until group theory has placed its seal of completeness on any particular construction. In an essay on the work of Laguerre,[11] Poincaré alludes to its novel non-Cartesian character. When Laguerre published his first paper in 1853, analytical geometry, according to Poincaré,

> was making itself over . . . by means of a revolution in some sense the opposite of the Cartesian reform. Before Descartes, chance alone, or genius, made possible the solution of a geometric problem. After Descartes we were in possession of infallible rules for obtaining results. To be a geometer it was enough to be patient. But a purely mechanical method that asks nothing of the mind by way of invention cannot be really fruitful. A new reform was therefore necessary. Poncelet and Chasles were its initiators. Thanks to them, we no longer need rely on either tireless patience or a stroke of good fortune for the solution to a problem, but rather on profound knowledge of the mathematical facts and their intimate relations.[12]

The method of Poncelet, Chasles, and Laguerre was thus a method of discovery more than a method of solution. It was clearly synthetic in nature and, as Poincaré points out, op-

[11] Edmond Laguerre (1834–1886), French mathematician.
[12] Poincaré, *Savants et ecrivains*, p. 86.

posite in intent to the Cartesian reform. Thus in some respects it marks the end of Cartesian thought in mathematics.

V

When one truly appreciates how far modern mathematics has surpassed the primitive science of spatial measurement and how much the "science of relations" has grown in recent years, it becomes clear that mathematical physics is each day opening new avenues of scientific objectification. Only after we have realized this does the stylized nature that scientists, with the aid of mathematics, create in the laboratory begin to seem less opaque than the nature that offers itself immediately to our observation. Conversely, once objective thought has been educated through the study of an organic nature of some sort, it reveals itself to be remarkably deep, for the simple reason that it is perfectible, correctable, and suggestive of new theories. The best way for a thinking subject to deepen its thoughts is still by meditating upon an object (as Decartes by his hearth meditated, in the *Discourse on Method*, upon his piece of wax — trans.). But rather than follow the example of the metaphysician who sits down at his hearth, one would do well to follow the mathematician who heads for his laboratory. Before long, every physics and chemistry laboratory may bear over its entrance the Platonic caveat, Let no one enter here who is not a geometer.

By way of illustration, let us compare Descartes's observation of his piece of wax with what I shall call the wax-drop experiment (an imaginary experiment that Bachelard invents to illustrate the use of careful modern techniques — trans.).

For Descartes the ball of wax was a clear symbol of the fleeting character of material properties. Simply by placing it near the fire he was able to alter or transform its consistency, its shape, its color, its feel, and its smell — an experiment that to us may seem rather crude but to Descartes proved the ambiguity of so-called objective qualities. Here is a lesson in doubt. Its point is to alienate the mind from experimental knowledge of substances, which are harder to know, Descartes tells us, than the soul. If the understanding were unable to find within itself the basis for the science of extended substances, then the substance of the ball of wax would evaporate along with the reveries of the imagination. It is only *intelligible* extension that subsists in the wax, since even its size is subject to increase or decrease depending on the circumstances. Descartes' refusal to base thought on experience is in fact final, even though he does ultimately return to the study of extended substance. From the first, however, he rules out any possibility of what I shall call progressive experimentation, any means of classifying or measuring the diversity of what is observed, any way of fixing the variables of the phenomenon in order to distinguish one from another. Descartes's desire was to apprehend directly the object's simplicity, unity, and constancy, and at the first sign of failure he was plunged immediately into doubt of *everything*. He failed to see the coordinating possibilities in directed experimentation and did not recognize how theory combined with experiment might restore the organic, and hence entire and complete, character of the phenomenon. What is more, by refusing to submit docilely to the lessons of experience, he condemned himself to overlook the fact that the variability of objective observation is immediately reflected in a corresponding mobility of subjective experience. If the wax changes, I change; I change with my sensation, which is, in the moment I conceive of it, the entire content of my thought; for to feel is to think in the broad sense that Descartes attaches to the *cogito*. But Descartes has a secret confidence in the reality of the soul as substance. Dazzled by the sudden light of the *cogito*, he never doubts the permanence of the *I* that is the subject of *I think*. Why is it the same being who feels first the hard wax and then the soft wax, when it is not the same wax that is felt on the two occasions? If the *cogito* were recast in the passive voice as *cogitatur ergo est*, would the active subject vanish along with the inconstancy and vagueness of its impressions?

This Cartesian partiality in favor of subjective experience will be all the more evident, perhaps, when we learn to bring more fervor to objective scientific experiment, to measure precisely the limits of our thoughts, and to match, in a strict and rigorous manner, thought to experiment, noumenon to phenomenon, rather than allow ourselves to be misled by the deceptive appearance of substances both subjective and objective.

Let us look, therefore, at how modern science goes about its business of progressive objectification. A modern physicist working with a ball of wax would not start with bees' wax straight from the hive but with chemically pure wax produced by careful purification techniques. The wax used is therefore in one sense a specific *moment* of a "method of objectification." It retains no trace of the fragrance of the flowers that entered into its composition, but it does bear the marks of the careful process of purification to which it was subjected. It is, in a manner of speaking, the product of artificial experience. Without an artificial experience of this kind, such a ball of wax — pure wax, not in its natural form — would never have come into existence.

The physicist would then melt this wax in a crucible and resolidify it in a slow, methodical manner. He can precisely control the rate of melting and solidification by using a small electric oven whose temperature can be regulated by adjusting the supply of electrical power. Thus the physicist gains *control of time*, the time during which the action of the heat affects the composition of the wax. In this way he can obtain a wax "droplet" whose shape and surface composition can be precisely controlled. Now that the "book of the microcosmos" is engraved, as it were, it remains to be read.

In order to study the surface of the wax, the physicist might expose it to a monochromatic beam of X-rays; he would do this in a very careful way, of course, and would never think of using "natural" white light, which in prescientific ages was thought to be of a simple nature. Thanks to the slow cooling of the ball of wax, the surface molecules will be oriented in a precise way relative to the surface of the drop. This orientation will determine the diffraction pattern of the X-rays and yield spectrograms similar to those obtained by Debye and Bragg for crystals. As is well known, crystal spectrograms, whose existence was predicted by von Laue, have given new life to the discipline of crystallography by enabling scientists to deduce the internal structure of various crystals. Similarly, our study of the wax drop may give us new knowledge of the surface structure of this form of matter. This new way of "reading" matter can be highly instructive. As Trillat points out, "orientation phenomena . . . are responsible for many surface properties, such as capillarity, oiliness, adherence, adsorption, and catalysis."[13] The outer film of a substance determines its relations with the outside world, a whole new realm for physicial chemistry to explore. It is by attending to this new realm that the metaphysician can best understand the influence of structure. We can examine the orientation of the molecule at various depths below the surface of the wax droplet. It turns out that the orientation gradually disappears as we go deeper and deeper into the interior; the microcrystals become less and less sensitive to the surface action, and beyond a certain depth there is complete statistical disorder. In the orientation zone, however, we observe an interesting set of phenomena having to do with the discontinuity of the molecular fields at the separation surface — the zone of material dialectic, as it were. In this intermediate region, various interesting experiments reveal an interplay between the physical and chemical properties of the wax and enable the physicist to alter its *chemical nature*. Trillat, for example, has reported on experiments pertaining to the stretching of colloidal gels. Using purely mechanical means to stretch the gel can cause marked differences in the X-ray diffraction patterns. Trillat's conclusion is as follows: "This pertains to the mechanical properties (of the gel) and also to the adsorption of dyes, according as the matter is oriented by traction or not: this may be an unsuspected way of affecting chemical activity."[14]

To affect chemical activity by mechanical means is in certain respects in keeping with the Cartesian ideal. But here

[13] Jean Trillat, "Etude au moyen des rayons X des phénomènes d'orientation moléculaire dans les composés organiques," *Activation et structure des molecules*, 1928, p. 461.
[14] Ibid., p. 456.

the artificial, constructive intent of the experiment, its impulse to greater and greater complexity, is so clear that it can only be regarded as yet another proof of the way science has extended our possibilities of experience, and as yet another instance of the non-Cartesian dialectic.

Is it even clear that crystallization can take place in the absence of ambient fields? The idea that crystallization is essentially the result of internal forces stemming from the substance itself and that it is possible to neglect outside influences is based on realist presuppositions. Indeed, it is striking to discover that surface crystallization depends on field discontinuities to such an extent that one can speak of substances that are superficially crystallized in the direction perpendicular to the surface while remaining amorphous in the direction parallel to the surface. The resulting structure is rather grasslike in structure, with the "blades" of grass implanted in the amorphous body of the substance in a definite manner. This new type of crystalline "growth" has already yielded considerable information about molecular structures.[15]

Anyone who is willing to recognize the importance of the new techniques, hypotheses, and mathematical theories that are involved in our proposed study of the drop of wax cannot fail to acknowledge that Descartes's metaphysical critiques have lost their edge. What is fleeting is not, as Descartes thought, the properties of the wax but the haphazard circumstances surrounding his observation of it; modern science coordinates its observations in its search for the qualities of matter. In nature the conditions of observation are *con-*

[15] See Jean Thibaud, "Etudes aux rayons X du polymorphisme des acides gras," *Activation et structure des molecules*, pp. 410ff.

fused, and all one has to do is put some order into the process in order to bring organization to the real. For science, then, the qualities of reality are functions of our rational methods. In order to establish a scientific fact, it is necessary to implement a coherent technique. Scientific work is essentially complex. Science is a discipline of active empiricism, which, rather than rely on whatever clear truths happen to lie ready to hand, actively seek its complex truths by artificial means. Innate truths naturally have no place in science. Reason has to be shaped in the same way as experience.

"Objective meditation" in the laboratory commits us to a path of progressive objectification that gives reality to both a new form of experience and a new form of thought. Subjective meditation is bent on attaining clear and definitive knowledge; objective meditation differs from this by the very fact that it makes progress, by its intrinsic need always to go further, to extend the limits of the known. The scientist, when he has done with his days' objective meditation, has his program of research for the following day in hand, and at the end of each working day he repeats the following article of faith: Tomorrow I shall know the truth.

VI

Looking now at the problem of scientific innovation from the psychological standpoint, it seems certain that the revolutionary character of modern science will have profound effects on the structure of the scientific spirit. Now, to say that the structure of the scientific spirit changes is just another way of saying that knowledge has a history. Human history, with all its passions and prejudices and its dependence on impulses of the moment, may well be a theater of eternal recurrence. But as history moves forward, there are some ideas that are not simply repeated; these are ideas that have been rectified, enlarged, completed, ideas that have definitively outstripped the limited and shaky principles on which they may once have been based. Now, the scientific spirit is essentially a way of rectifying knowledge, a way of broadening the horizon of what is known. Sitting in judgment, it condemns its historic past. Its structure is its awareness of its historical errors. For science, truth is nothing other than a historical corrective to a persistent error, and experience is a corrective for common and primary illusions. The intellectual life of science depends dialectically on this differential of knowledge at the frontier of the unknown. The very essence of reflection is to understand that one did not understand before. The non-Baconian, non-Euclidean, and non-Cartesian philosophies are historical dialectics that grew out of the correction of an error, the extension of a system, or the completion of an idea.

In order for the new scientific spirit to take on the same formative value as a new economic policy,[16] all that is needed is a little social life, a little human sympathy. For many scientists, who passionately lead the dispassionate life, the resolution of today's scientific problems will determine the future of reason itself. Reichenbach has spoken, rightly I think, of a generational conflict over the deep meaning of science.[17] While visiting J. J. Thomson at Cambridge, Karl Compton happened to meet G. P. Thomson, the elderly

[16] The French makes clear that Bachelard is here alluding to the Soviet Union's New Economic Policy. — Trans.
[17] Reichenbach, *La philosophie scientifique*, pp. 23–24.

physicist's son, who had come up for the weekend. The three men spent some time examining photographs taken by means of electron waves, about which Compton made the following remark: "It was truly a dramatic event to see the grand old man of science, who had spent the best years of his life arguing that the electron is a particle, full of enthusiasm for the work of his son, which revealed that electrons in motion are in fact waves."[18] The distance traveled from father to son is a measure of the philosophical revolution entailed by the abandonment of the notion that the electron is a thing. The intellectual courage required for such a revision in our realist principles commands our admiration. Physicists have been obliged, three or four times in the past twenty years, not simply to change their minds but, intellectually speaking, to make a totally fresh start.

When, moreover, we realize what an incomplete state modern science is in, we begin to gain some intimate idea of the meaning of "open-minded rationalism" (*le rationalisme ouvert*). To be rational and yet open-minded is to experience genuine surprise at the implications of theoretical speculation. Juvet puts it quite well: "The surprise created by a new idea or association of ideas is surely the most important element in the progress of the physical sciences, for it is astonishment that excites logic, which is always rather cold, and that forces scientists to make new connections. But the ultimate cause of progress, the reason for our surprise itself, has to be sought in the force fields that new associations of ideas set up in our minds, fields whose strength measures

[18] Hans Reichenbach, *Scientifische Monatschrift*, 1929, vol. 28, p. 301. Cited by Haissinsky, p. 348.

the good fortune of the scientists lucky enough to bring those ideas together."[19]

Confronted with the surprising principles of the new quantum mechanics, even Emile Meyerson, who expended such vast quantities of meditation and erudition to prove that relativity was in fact classical physics, has suddenly been gripped by doubt. There is reason to doubt that anyone will ever write a *Quantum Deduction* to complete the proof that Meyerson began in his *Relativistic Deduction*. As he himself confesses, "compared with the theories I have examined in my books, quantum theory admittedly occupies a place apart, and it does not seem possible, in my view, to attempt for it what I believe I have accomplished for the theory of relativity."[20] For Meyerson, quantum theory is essentially an aberration, and he is not far from claiming that the "arithmetization of the possible" is really an irrational doctrine. My view is quite the contrary: It is that the quantum theory extends, in a positive sense, our conception of the real and marks a triumph of the new reason over irrationalism. The mind must be made ready to receive the quantum idea, and this can be done only by systematically expanding the scientific spirit.

I further believe that relativity already marks the triumph of an eminently inductive theory, and that the success of efforts to deduce, for pedagogical purposes, certain consequences of relativity theory in no way diminishes the brilliance of Einstein's achievement or makes it any less surprising. The genius of de Broglie in founding wave mechanics and of Heisenberg in founding matrix mechanics has been

[19] Gustave Juvet, *La structure* (Paris: F. Alcan, 1933), p. 105.
[20] Meyerson, *Le cheminement de la pensée*, vol. 1, p. 67.

no less astonishing and no less historically unprecedented. These discoveries have relegated both classical and relativistic mechanics to the past, as two more or less crude approximations to more subtle and complete theories.

Will some still more general theory swallow up all these astonishing advances and establish itself as the immutable truth? Will yet another advance bring order to chaos and rule the universe? Doubtless it is Meyerson's profound wish that such will prove to be the case. When Meyerson shows how modes of thought endure for centuries and demonstrates the persistence of primitive ways of thinking in the most modern of minds, he draws the conclusion that the brain cannot evolve any more rapidly than any other organ. This argument of Meyerson's is obviously the argument of caution, and it would be hard to gainsay him without venturing the riskiest of speculations. Still, I shall try. For isn't the brain the true center of human evolution, the terminal bud of the vital spirit? With its manifold connections, is it not the organ of innumerable possibilities? When Juvet suggestively alludes to the "force fields" created in the mind by bringing together two different ideas, he encourages us to interpret the traditional association of ideas in a more dynamic light and to give to Fouillée's notion of an idea force an almost physical interpretation. An evolving idea is an organic center that swells to greater and greater proportions. A static brain would be a brain that never drew any new conclusions. In order to prove the continuity of the mind, are there no better arguments than commonplace, effortless thoughts, the thoughts that control our muscles and by so doing merge with what no longer evolves? If so, then everything is already complete: the soul, the body, the world itself, the world as it is given to us with its grand and noble features. The philosophers, for their part, hold out to us the idea of communion with an all-enveloping reality, to which the scientist can hope for nothing better than to return, as to a philosophy original and true. But if we really want to understand our intellectual evolution, wouldn't we do better instead to pay heed to the anxiety of thought, to its quest for an object, to its search for dialectical opportunities to escape from itself, for opportunities to burst free of its own limits? In a word, wouldn't we do better to focus on thought in the process of objectification? For if we do, we can hardly fail to conclude that such thought is creative.

The psychological advance brought about by mathematical physics has been described by Juvet. He points out that the boldest and most fruitful ideas have been the work of very young scientists: "Heisenberg and his pupil Jordan were born with the century. In England, an astonishing genius . . . Dirac developed a new and original method and discovered the deep theoretical reasons for what has been called the spin of the electron; he was only twenty-five years old at the time. If we recall also that Bohr was very young in 1913 when he proposed his model of the atom, and that Einstein was twenty-five when he discovered special relativity and shortly thereafter proposed for the first time an explanation of the laws of radiation for quanta of light . . . it seems reasonable to conclude that the twentieth century has witnessed a mutation in man's brain or mind of a sort apt to help him unravel the laws of nature, much as the precocity of the Abels, Jacobis, Galois, and Hermites of the previous century may have been due to a mutation of the mind apt to further the understanding of mathematical objects."[21]

Each of us can relive these intellectual mutations by thinking back to the turmoil and emotion that the new doctrines caused us personally when we first learned of them. They required such effort of learning that they did not seem natural. But *natura naturans* (to use Spinoza's term — trans.) is at work in our very souls; at some point we each realized that we had understood. By what light do we recognize the importance of these sudden syntheses? By an ineffable light that brings security and happiness to our minds. This intellectual happiness is the first sign of progress. This is the moment to recall, with Jean Hering, the phenomenologist, "that the most advanced person will always be able, thanks to his broader horizon, to understand his inferiors . . . whereas the reverse is impossible.[22] Understanding has a dynamic dimension; it is a spiritual élan, a vital élan. Einsteinian mechanics added to our understanding of Newtonian concepts. De Broglie's mechanics is adding to our comprehension of the concepts of classical optics and mechanics, which the new physics is molding into a new synthesis that extends and completes the epistemology of Descartes. If only we were capable of immersing ourselves in scientific research with all our strength and of studying our psychological development as we study other aspects of our cultural history, we would be able to feel the sudden animation that has been given to the soul by the creative syntheses of mathematical physics.

[21] Juvet, *La structure*, p. 134.
[22] Jean Hering, *Phénoménologie et philosophie religieuses* (Strasbourg, 1925), p. 126.

Index

Gaston Bachelard was born in 1884 in Bar-sur-Aube, France, the son—and grandson—of a cobbler. When he died in 1962, he was a member of the Institute and one of the most illustrious names in French philosophy. He was the author of numerous books, among which are *The Poetics of Reverie, The Poetics of Space,* and *The Psychoanalysis of Fire*—all published by Beacon Press.